Three

lit

CONTROL PROCESSES IN FISH PHYSIOLOGY

Control Processes in Fish Physiology

Edited by J.C. Rankin, T.J. Pitcher and R.T. Duggan

CROOM HELM
London & Canberra

British Library Cataloguing in Publication Data

Control processes in fish physiology.
1. Fishes – Physiology
I. Rankin, J.C. II. Pitcher, Tony J.
III. Duggan, R.T.
597.01   QL639.1
ISBN 0-7099-2246-9

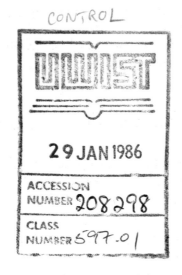
Typeset by Leaper & Gard, Bristol
Printed and bound in Great Britain

CONTENTS

DEDICATION

To Professor J.M. Dodd, F.R.S., F.R.S.E.

PREFACE

The 3rd Symposium on Fish Physiology held in Bangor, North Wales, in September 1981 drew together fish physiologists from many countries. After discussion at the symposium, it was felt that, rather than publish the proceedings of the meeting as a whole, it would be of greater value to ask selected speakers to summarise from a wider perspective how some of the major processes of fish physiology are controlled. This book therefore provides a useful introduction to the senior undergraduate and postgraduate student of zoology or biology, as well as providing reviews for the specialist physiologist and those with an interest in applied fish biology.

Control may be exerted at a number of different levels in animals. For example, control of the expression of genetic information governs differentiation of cells and tissues: enzymes are controlled at the biochemical level. This book concentrates on two further levels; the control of the functioning and co-ordination of organ systems, and the control of activity and behaviour of the whole fish. The physiological processes covered in the book are all subject to a combination of endocrine and nervous control.

The book starts with a survey of the role of the autonomic nervous system in fish and proceeds to discuss the regulation of the alimentary canal, where the autonomic system exerts much control. In Chapter 3 respiration and circulation are considered, which both involve endocrine and autonomic regulation. This is followed by the kidney, in which nervous control has been widely, and perhaps incorrectly, assumed to be relatively unimportant compared to the effects of circulating hormones. Chapter 5 ends this section of the book with a consideration of pancreative control of metabolism, an example of a purely endocrine control mechanism.

The next four chapters are devoted to endocrine and nervous control of fish behaviour, from reproductive homing to agonistic behaviour and spawning. Integration of control by the central nervous system also receives a broad treatment. Chapters 6-8 aim to be comprehensible to the student of fish behaviour who wishes to read a synoptic account of how hormones affect the behaviours measured in behavioural experiments.

Chapters 10-12 provide an up-to-date review of the environmental and hormonal control of reproduction in teleosts and elasmobranchs, with particular regard to applications in fish culture.

The original Bangor Symposium was generously supported by the Royal Society, the Society for Endocrinology, the University of Wales and the University College of North Wales. The session on reproductive physiology held at the Bangor Symposium was organised in honour of Professor J.M. Dodd FRS, FRSE, on his retirement from the Lloyd Roberts Chair of Zoology at the University College of North Wales. We feel that it is fitting that this book be dedicated to Professor Dodd, not only in respect of his notable contributions to fish endocrinology, but also because it reflects the concern which he has shown for zoology teaching throughout his distinguished career. The book is aptly concluded by a major review of the control of elasmobranch reproduction authored by Professor Dodd, his wife and co-worker for many years, Dr Margaret Dodd, and Dr R.T. Duggan.

J.C. Rankin,
T.J. Pitcher,
R.T. Duggan.

CONTRIBUTORS

J.A. Brown, Lecturer, Department of Zoology, University of Hull, Hull HU6 7RX, England

P.J. Butler, Senior Lecturer, Department of Zoology and Comparative Physiology, University of Birmingham, PO Box 363, Birmingham B15 2TT, England

J.M. Dodd, Professor Emeritus, School of Animal Biology, University College of North Wales, Bangor, Gwynedd LL57 2UW, Wales

M.H.I. Dodd, Senior Postdoctoral Research Fellow, School of Animal Biology, University College of North Wales, Bangor, Gwynedd LL57 2UW, Wales

R.T. Duggan, Senior Biochemist, Research and Development Laboratories, RIA (UK) Ltd., Armstrong Estate, Washington, Tyne and Wear NE37 1PP, England

R. Fänge, Professor, Department of Zoophysiology, University of Göteborg, Box 250 59, S 400, 31 Göteborg, Sweden

D.J. Fletcher, Postdoctoral Research Fellow, Marine Science Laboratories, University College of North Wales, Menai Bridge, Gwynedd LL59 5EH, Wales

D.J. Grove, Senior Lecturer, Marine Science Laboratories, University College of North Wales, Menai Bridge, Gwynedd LL59 5EH, Wales

D.M. Guthrie, Professor, Department of Zoology, University of Manchester, Oxford Road, Manchester M13 9PL, England

A.D. Hasler, Professor Emeritus, Laboratory of Limnology, University of Wisconsin, Madison, Wisconsin 53706, USA

I.W. Henderson, Senior Lecturer, Department of Zoology, University of Sheffield, Sheffield S10 2TN, England

S. Holmgren, Docent, Department of Zoophysiology, University of Göteborg, Box 250 59, S-400, 31 Göteborg, Sweden

B.W. Ince, Lecturer, Department of Biological Sciences, University of Aston, Birmingham B4 7ET, England

J.D. Metcalfe, Postdoctoral Research Fellow, Department of Zoology and Comparative Physiology, University of Birmingham, PO Box 363, Birmingham B15 2TT, England

Contributors

A.D. Munro, Postdoctoral Research Fellow, School of Animal Biology, University College of North Wales, Bangor, Gwynedd LL57 2UW, Wales

S. Nilsson, Docent, Department of Zoophysiology, University of Göteborg, Box 250 59, S-400, 31 Göteborg, Sweden

T.J. Pitcher, Lecturer, School of Animal Biology, University College of North Wales, Bangor, Gwynedd LL57 2UW, Wales

J.C. Rankin, Lecturer, School of Animal Biology, University College of North Wales, Bangor, Gwynedd LL57 2UW, Wales

A.P. Scott, Senior Scientific Officer, Ministry of Agriculture, Fisheries and Food, Fisheries Laboratory, Lowestoft, Suffolk NR33 0HT, England

N.E. Stacey, Assistant Professor, Department of Zoology, University of Alberta, Edmonton, Alberta, Canada T6G 2E9

J.P. Sumpter, Lecturer, Department of Applied Biology, School of Biological Sciences, Brunel University, Uxbridge, Middlesex UB8 3PH, England

V. deVlaming, Visiting Assistant Professor, Department of Animal Physiology and the Aquaculture Program, University of California, Davis, California 95616 USA

1 AUTONOMIC NERVE FUNCTIONS IN FISH

Stefan Nilsson, Susanne Holmgren and Ragnar Fänge

Autonomic nerves control the function of most organs in fish, as in higher vertebrates. The aperture of the iris, blood pressure, blood flow through the gills for oxygenation, blood delivery to different parts of the body, heart performance, gastric motility, swimbladder functions, colour change and release of catecholamines from chromaffin tissue are examples of mechanisms influenced by the activity of autonomic nerves. The aim of this chapter is to give a brief summary of the available information on the anatomy and physiology of the autonomic nervous system in cyclostomes, elasmobranchs, teleosts and dipnoans, pointing out some differences and similarities between the fish groups, and between fish and higher vertebrates. In this short review we have in many cases chosen to give examples from our own experience when several studies indicate the same result. More extensive reviews of the autonomic innervation in fish have been made by e.g. Burnstock (1969), Campbell (1970b), Santer (1977), Holmgren and Nilsson (1982), Nilsson (1983).

The action of an autonomic nerve fibre on its target organ is mediated by a chemical transmitter substance, released from the autonomic nerve and acting on specific receptors in the effector organ. Acetylcholine and adrenaline/noradrenaline were early identified as transmitters in the autonomic nervous sytem (Loewi and Navratil 1926). Recently a number of other substances have been suggested as 'non-adrenergic, non-cholinergic' (NANC) transmitters in autonomic nerves in mammals. Amongst these are ATP (Burnstock 1981), 5-hydroxytryptamine (Gershon 1981), certain amino acids and about 20 'neuropeptides', e.g. vasoactive intestinal polypeptide (VIP), substance P, neurotensin, somatostatin and gastrin (Furness and Costa 1980). The presence of some of the latter, e.g. VIP, substance P, enkephalin and 5-hydroxytryptamine, in fish autonomic nerves has been suggested from histochemical evidence (Langer, Van Noorden, Polak and Pearce 1979; Watson 1979; Holmgren, Vaillant and Dimaline 1981). Most of these NANC transmitters have been found in the enteric nervous system of the gut, and will therefore be discussed in Chapter 2.

1

Adrenergic Transmission

The predominant catecholamine in mammalian adrenergic neurons is noradrenaline. Measurements of catecholamine content in different organs of elasmobranchs suggest that this is also the case in elasmobranch adrenergic nerves (von Euler and Fänge 1961; Stabrovski 1969). On the other hand, the adrenergic neurons in teleost fish, like those in amphibians, store both adrenaline and noradrenaline with a predominance of adrenaline. The catecholamines are probably stored together in the same neuron, and may be released together, both acting as transmitter substances. Experiments with *Gadus*, also show that the adrenaline is synthesized intraneuronally from noradrenaline by the enzyme phenylethanol-amine-N-methyl transferase (PNMT; Abrahamsson and Nilsson 1976).

The adrenergic transmitter acts on adrenoceptors (adrenergic receptors) of either the alpha or beta type in the effector organs (Ahlqvist 1948). Quantitative studies of alpha-adrenoceptors in fish have been made, e.g. in melanophores from *Phoxinus*, where they mediate pigment aggregation, and in *Gadus* and *Salmo* coeliac arteries, *Gadus* spleen, *Ctenolabrus* swimbladder and *Gadus* iris sphincter (Holmgren and Nilsson 1982; Nilsson 1983), where the receptors in all cases mediate a constriction of smooth muscle. In these studies the alpha-adrenoceptors show minor differences in sensitivity to adrenergic agonists and antagonists. This may be due to true variations in receptor properties between different organs and different fish species, but also to dissimilarities in access to the receptor site and inactivation mechanisms for the transmitters in the various organs. Some of the results also suggest smaller variations in properties between mammalian and fish alpha-adrenoceptors (Holmgren and Nilsson 1982).

Beta-adrenoceptors are found in the heart, urinary bladder, swimbladder and gills (Holmgren and Nilsson 1982; Nilsson 1983) in different species of fish. Stimulation of the beta-adrenoceptors of the heart causes an increased activity of the heart, while in the other tissues an interaction between catecholamines and the beta-adrenoceptors causes an inhibition (relaxation) of the smooth muscles. In mammals the beta-adrenoceptors have been subdivided into two groups, $beta_1$ and $beta_2$, using pharmacological criteria. Few studies in fish have aimed to characterise the beta-adrenoceptors However, the beta-adrenoceptors of *Salmo* heart are suggested to be of the $beta_2$ subtype, in contrast to mammals where both $beta_1$ and beta$_2$ adrenoceptors are present in the heart, but beta$_1$ receptors dominate (Ask, Stene-Larson and Helle 1980).

Cholinergic Transmission

Acetylcholine is the transmitter in autonomic ganglia, released from preganlionic nerve terminals synapsing on postganglionic neurons. The receptors of the postganglionic neurons are of the nicotinic type. This is shown in *Gadus*, where the transmission through the coeliac ganglion and the vesicular ganglion is blocked by the ganglion-blocking agents hexamethonium or mecamylamine, which act mainly on nicotinic receptors. Acetylcholine released from postganglionic cholinergic nerve endings acts on muscarinic receptors in the effector organ in teleosts, elasmobranchs and dipnoans, as in all higher vertebrates. In lampreys the cholinergic receptors on the heart instead appear to be of the nicotinic type (Lukomskaya and Michelson 1972).

General Anatomy

The autonomic nervous system comprises all efferent nervous pathways, with a ganglionic synapse outside the central nervous system (Campbell 1970a). Classically the mammalian autonomic nervous system has been subdivided into a parasympathetic, a sympathetic and an enteric part (Langley 1921). Parasympathetic pathways are characterised by a cranial or sacral outflow, with long preganglionic and short postganglionic neurons, and acetylcholine was originally thought to be the only postganglionic transmitter. Sympathetic pathways have a thoracic or lumbar spinal outflow, with short preganglionic neurons connecting to postganglionic neurons in the sympathetic chains. Noradrenaline or adrenaline are the transmitters in most postganglionic sympathetic neurons. The enteric neurons are those intrinsic to the gut, involved in the control of gut motility, blood flow and secretion from mucosal cells in stomach and intestine.

This very simplified picture, based on a combination of anatomical and functional criteria of the autonomic nervous system in mammals is not valid in non-mammalian vertebrates (Campbell 1970b). We therefore prefer to use the more general, solely anatomical, terminology proposed by Nilsson (1983), where the *cranial autonomic system* includes all autonomic pathways with a cranial outflow from the central nervous system, the *spinal autonomic system* includes all pathways with a spinal outflow and the *enteric system*, as above, consists of neurons intrinsic to the gut.

Generally, the *cranial outflow* of autonomic pathways reaches effector organs in the head and, via the intestinal branches of the vagi, the anterior part of the trunk. Preganglionic myelinated fibres synapse on postganglionic neurons in ganglia close to the effector organ. Postganglionic, unmyelinated fibres for instance innervate the iris, gills, heart and stomach. The *spinal autonomic outflow* of myelinated preganglionic fibres make synaptic connections with postganglionic neurons in the sympathetic chains, or pass through these chains to make synaptic connections in collateral ganglia in the trunk. The postganglionic, unmyelinated neurons reach effector organs in the whole body, including iris, heart, vasculature, stomach, intestine, chromatophores, swimbladder (lung), gallbladder, urinary bladder and spleen.

Cyclostomes

The autonomic nervous system in the cyclostomes is poorly developed compared to the other vertebrates, and it has often been difficult to determine whether fibres in the peripheral nerves have been autonomic or sensory.

The cranial outflow in all cyclostomes includes the vagus nerve (X). In myxinoids this seems to be the only pathway for cranial autonomic fibres, while in lampetroids autonomic fibres to the gill region run in the facial (VII) and glossopharyngeal (IV) nerves (Nicol 1952). The left and right vagi unite, forming the *nervus intestinalis impar*. A vagal innervation of the heart and gut is described in lampetroids (Nicol 1952), while a well-developed vagal control is described only for the gallbladder in myxinoids (Fänge and Johnels 1958). The function of the vagal fibres reaching the myxinoid intestine is uncertain, and the heart lacks an innervation.

Spinal autonomic fibres leave the central nervous system in the spinal nerves. No sympathetic chains exist, and scattered clusters of ganglion cells may represent the autonomic ganglia (Nicol 1952). The spinal autonomic fibres reach most inner organs in the lampetroids (Nicol 1952; Johneis 1956). Fluorescence histo-chemistry has shown that some of these fibres contain catechol-amines and may thus be adrenergic. In *Myxine*, the visceral fibres of spinal autonomic origin are probably vasomotor, and it is doubtful if any fibres reach the gut smooth muscles. In both cyclostome groups spinal fibres form an extensive subcutaneous nerve plexus with possible involvement in colour changes, mucus release or cutaneous blood flow (Nicol 1952; Fänge, Johnels and Enger 1963; Campbell 1970b).

Elasmobranchs

Cranial autonomic fibres are found in the oculomotor (III), facial (VII), glossopharyngeal (IX) and vagus (X) nerves in elasmobranchs. The fibres innervate the eye (III), blood vessels of head and anterior trunk region, pharyngeal smooth muscle (VII, IX, X) and possibly the gill arches (IX, X). Vagal fibres also reach the heart, the stomach and the anterior part of the spiral intestine (Young 1933; Nicol 1952).

The spinal autonomic outflow from the central nervous system reaches segmentally arranged paravertebral ganglia. Some of these are connected longitudinally and with the contralateral paravertebral ganglia, but no proper sympathetic chains comparable to those in higher vertebrates exist. The most anterior paravertebral ganglia (gastric ganglia) are associated with large clusters of catecholamine-storing chromaffin cells, forming the two axillary bodies (Leydig 1853). An anterior splanchnic nerve leaves the axillary body on either side. The two nerves join to form a plexus along the coeliac artery reaching the gut and the liver (Young 1933; Nicol 1952). An anastomosis between the anterior splanchnic nerve and the vagus has been found in some elasmobranchs (Young 1980a).

Teleosts

In teleosts cranial automomic fibres are found only in the oculomotor (III) and the vagus (X) nerves. The preganglionic fibres from the oculomotor nerve synapse in the ciliary ganglion with postganglionic neurons, which reach the eye via the ciliary nerves. Vagal fibres run in the branchial, cardiac, and intestinal branches to gills, heart and viscera (pharynx, oesophagus, stomach and swimbladder) respectively (Stannius 1849; Young, 1931).

The preganglionic fibres in the teleost spinal outflow from the central nervous system leave the spinal nerves in white rami communicantes and enter the segmentally arranged ganglia. In the ganglia some fibres synapse with the ganglion cells, while other fibres pass through to make synaptic connections with the postganglionic neurons closer to the effector organ. Fibres leaving the ganglia either re-enter the spinal nerves via grey rami communicantes, or form autonomic nerves (anterior and posterior splanchnic nerves) passing directly to the visceral organs (Stannius 1849; Young 1931; Figure 1.1).

The segmentally arranged paravertebral ganglia ('sympathetic ganglia') are longitudinally connected to form well-developed sympathetic chains. In teleosts these chains continue into the head region and carry ganglia in contact with the cranial nerves V + VII, IX and X. Postganglionic fibres are sent off to the respective cranial nerves (via grey rami communicantes) and follow these, e.g. to melanophores and blood vessels. A large number of fibres join the vagi, forming 'vago-sympathetic' trunks, which thus carry both cranial and spinal autonomic fibres to gills, heart, stomach and swimbladder (Stannius 1849; Young 1931; Nilsson 1976).

The first two or three ganglia in the spinal part of the right sympathetic chain are fused, forming the elongated coeliac ganglion, which sends postganglionic fibres in the anterior splanchnic nerve to the visceral organs. Fibres from the left chain reach the coeliac ganglion and anterior spanchnic nerve via a commissure. Other commissures (number depending on teleost species) may further interconnect the two chains (Stannius 1849; Young 1931; Nilsson 1983).

A posterior splanchnic nerve (vesicular nerve) leaves the posterior part of the sympathetic chains and innervates the urogenital organs and rectum (Young 1931; Burnstock 1969; Figure 1.1).

Dipnoans

Cranial autonomic fibres are found only in the vagus (X) of *Protopterus* and *Lepidosiren*, while in *Neoceratodus* fibres may also be present in the oculomotor (III) nerve (Nicol 1952). The vagal autonomic fibres reach the gut, lung and heart (Jenkin 1928).

The sympathetic chains are poorly developed, with small segmental ganglion clusters and very thin longitudinal interganglionic connections. The most anterior part is associated with the vagus, and fibres may be sent off forming a vagosympathetic trunk. No further extensions of the sympathetic chains into the head region as in teleosts have been found (Jenkin 1928).

Chromaffin Tissue

Mammalian chromaffin cells are defined as cells with the following characteristics:

Figure 1.1: Simplified diagram showing the arrangement of the paravertebral ganglion chains (sympathetic chains) in a teleost, the cod (*Gadus morhua*). Legend: ant spl, anterior splanchnic nerve(s); cil g, ciliary ganglion; cil nn, long and short ciliary nerves; coel g, coeliac ganglion; comm, commissure between left and right sympathetic chain; int X, intestinal branch of vagus nerve; post spl, posterior splanchnic nerve (vesicular nerve); sat g, satellite ganglion. Roman numerals refer to the cranial nerves, while Arabic numerals indicate the spinal nerves. Note the nerves to the chromaffin tissue of the head kidney (*), and the nerves from the sympathetic ganglia which enter the cranial nerves (also along the more posterior parts of the intestinal branches of the vagi).

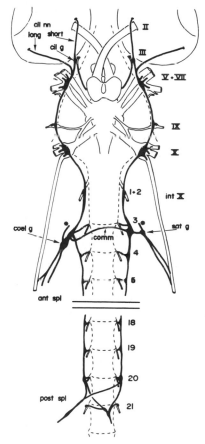

(1) They develop from the neuroectoderm,
(2) They are innervated by preganglionic fibres,
(3) They synthesize and release catecholamines, and
(4) They store catecholamines in sufficient amounts to give a chromaffin reaction (Coupland 1965).

All of these criteria have not always been determined for the catecholamine-rich cells called chromaffin in lower vertebrates, but the term chromaffin has still been used for convenience.

In mammals the majority of the chromaffin cells are situated in the adrenal medulla. Fish have no true adrenals, but clusters of chromaffin cells can be found in several organs, the location depending on species. The chromaffin tissue is often strategically placed to allow a rapid effect on a nearby organ by the catecholamines released from the chromaffin cells. For example, in many fish big clusters of chromaffin tissue are found in the large veins close to the heart, in the cyclostomes and dipnoans even within the heart itself. In fish, catecholamines from the chromaffin tissue probably are of crucial importance in the adrenergic control of different organs, often perhaps more important than the effect of an adrenergic innervation.

In the cyclostomes, which have a poorly developed autonomic nervous system, chromaffin cells are found in a number of tissues, like the large veins and arteries, the portal heart and the atrium and ventricle of the systemic heart. The chromaffin cells contain both adrenaline and noradrenaline (Euler and Fänge 1961). The cells of the heart are small, intensely fluorescing (after treatment for fluorescence histochemistry) cells with long processes containing granules (Bloom, Östlund, Euler, Lishajko, Ritzen and Adams-Ray 1961; Otsuka, Chihara, Sajurada and Karda 1977). Treatment of *Myxine* with reserpine, which will cause a depletion of catecholamines from the chromaffin cells, produces bradycardia or even heart arrest (Fänge and Östlund 1954).

The chromaffin tissue in elasmobranchs is associated with the paravertebral autonomic ganglia; the largest clusters of chromaffin cells form the axillary bodies together with the gastric ganglia (Leydig 1853). Catecholamine analysis shows that the axillary bodies contain more noradrenaline than adrenaline (Euler and Fänge 1961). Stimulation of the anterior spinal cord as well as 'stress' causes an increase of catecholamine content in the plasma of *Squalus*.

The head kidney (anterior part of the kidney) of teleosts contains a large number of chromaffin cells, often lining the posterior cardinal veins passing through the kidney (Nandi 1964). In addition small intensely fluorescent cells are found in several of the autonomic ganglia in *Gadus*. Stimulation of the bundle of myelinated nerve fibres running to the head kidney in *Gadus* causes a release of catecholamines, mainly adrenaline, from the chromaffin cells (Nilsson *et al.* 1976). This release can be blocked by hexamethonium, indicating the preganglionic nature of the fibres. The amount of catecholamines released from the head kidney during stimulation is big enough to produce a tachycardia and an increased flow through the gills in isolated preparations (Holmgren 1977; Wahlqvist 1981; Figure 1.2). 'Stress' produces an increase in plasma catecholamine content, which is not obtained if the outflow of spinal fibres to the chromaffin tissue is sectioned (Nilsson *et al.* 1976).

The distribution of chromaffin tissue in the dipnoan *Protopterus* is similar to that in cyclostomes. Numerous chromaffin cells are found in the atrium of the heart, and the intercostal arteries and the left cardinal vein are also lined by chromaffin tissue (Abrahamsson, Holmgren, Nilsson and Pettersson 1979a). Disturbance of the fish causes a massive release of catecholamines into the blood (Abrahamsson, Holmgren, Nilsson and Pettersson 1979a).

Autonomic Control of the Cardiovascular System

The fish circulatory system forms a single circuit with the heart, the branchial vasculature and the systemic vasculature coupled in series. The heart consists of a thin-walled sinus venosus, and a single atrium and ventricle pumping the blood via the bulbus cordis (in teleosts bulbus arteriosus) into the ventral aorta. In cyclostomes the pumping action of the branchial and caudal 'hearts' and the portal vein heart help to maintain the blood flow through the system.

Innervation of the Heart

The heart of myxinoid cyclostomes has no functioning innervation, and the heart is insensitive to adrenergic and cholinergic drugs (Fänge *et al.* 1963). Lampetroid cyclostomes have a vagal innervation of the heart, which in these fish is excitatory (!). Acetylcholine stimulates the heart, probably by an action on a nicotinic type of receptor. Less pronounced stimulation of the lampetroid heart can be obtained with adrenergic drugs (Lukomskaya and Michelson 1972). An adrenergic control of the cyclostome heart is most likely exerted by the chromaffin cells present in vast amounts in the heart tissues (see p. 8).

All gnathostomous fish examined have, like the tetrapods, an inhibitory vagal innervation of the heart. The preganglionic vagal fibres synapse with the cell bodies of postganglionic neurons in the walls of the sinus venosus or in the border region between the sinus venosus and the atrium. The postganglionic fibres produce a negative chronotropic effect on the elasmobranch heart. In teleosts, and probably also the dipnoans, vagal fibres innervate the pacemaker cells and atrial musculature, thus controlling both chronotropy and inotropy of the heart (Laurent 1962; Johansen and Burggren 1980; Figure 1.3).

The presence and importance of an excitatory adrenergic innervation of the fish heart varies with group and species (Figure 1.3). Cyclostomes and elasmobranchs lack an adrenergic innervation. In elasmobranchs a variation in the degree of cholinergic inhibitory tonus could be responsible for the total nervous regulation of the heart. Amongst teleosts all species investigated, with the exception of *Pleuronectes*, have been found to have an adrenergic innervation. The fibres are spinal in origin, but usually leave the sympthetic chain in the cranial region to join the vagi, which reach the heart as 'vago-sympathetic trunks'. In *Gadus* adrenergic fibres to the heart also run in the fused first two spinal nerves (Holmgren 1977), and in other fish they may reach the heart along coronary vessels (Gannon and Burnstock 1969).

Innervation of the Branchial Vasculature

Little is known about the innervation of the branchial vasculature in cyclostomes. Catecholamines produce a dilation followed by a constriction, which could be blocked by beta- and alpha-adrenergic antagonists respectively. Acetylcholine constricts the branchial vasculature (Reite 1969). Whether autonomic nerves can produce these effects in the intact animals is unknown.

The few studies made on the control of branchial vascular resistance in elasmobranchs by catecholamines demonstrate the presence of an alpha-adrenergic constrictor and a beta-adrenergic dilator mechanism in the gill vasculature (Davies and Rankin 1973). Adrenergic fibres are unlikely to be involved in the control; catecholamines released from chromaffin tissue are probably responsible for the control of branchial flow in elasmobranchs (Davies and Rankin 1973).

Figure 1.2: Concentration-response curves for catecholamines acting on the isolated perfused gill arch of the cod, *Gadus morhua*. The hatched area shows the range of plasma concentrations of adrenaline in resting (approx. 30 nmol) and 'stressed' cod (approx. 300 nmol). A, adrenaline; NA, noradrenaline.

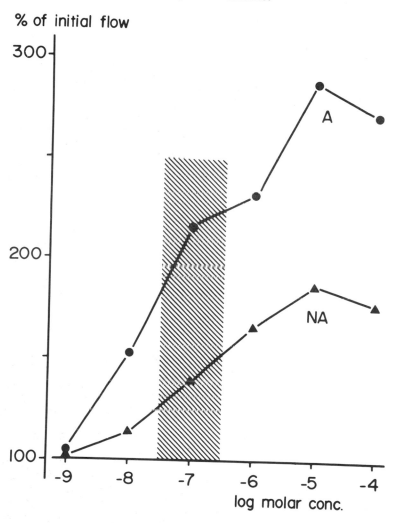

Figure 1.3: Inotropic effects if acetylcholine (ACh) and adrenaline (ADR) on isolated strips from the atrium of the cod (*Gadus morhua*) heart. The strips are paced electrically with pulses at 1 Hz, 1 ms duration and 10 V. The agonist drugs were added to give final bath concentrations of 10^{-8}, 3×10^{-8}. 10^{-7} ... 10^{-3} mol/litre (M).

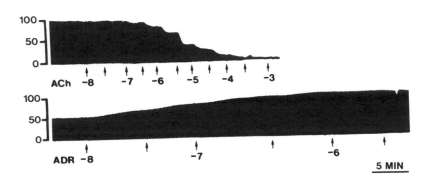

Reproduced with permission from Holmgren 1977

The control of blood flow in the teleost gills has been more extensively studied. In general, two main blood pathways through the gills can be discerned. The blood passes the secondary lamella of the gills where it is oxygenated, and is then either delivered to the different parts of the body via the dorsal aorta (arterio-arterial pathway), or passes into the venous system of the gills and back to the heart (arterio-venous pathway). Numerous experiments show that the total branchial vascular resistance is reduced by catecholamines acting via beta-adrenoceptors (for references see Wood 1975). The arterio-venous pathway is constricted by adrenaline acting via an alpha-adrenergic mechanism (Nilsson and Pettersson 1981). There are, however, few studies regarding the actual innervation of the branchial vasculature by autonomic nerves. In *Gadus* the gills are innervated by branchial nerves, which are branches of the glosso-pharyngeus and vagus. Cranial autonomic cholinergic fibres in the vagus constrict the arterio-arterial pathway, and adrenergic fibres from the sympathetic chain ganglia, which run in both vagal and glossopharyngeal branchial branches, dilate the arterio-arterial pathway and (especially) constrict the arteriovenous pathway (Nilsson and Pettersson 1981). The most important control of the total branchial vascular resistance may, however, be due to circulating catecholamines, while the major function of the adrenergic nerves may be to control the lamellar recruitment and the shunting between the arterio-arterial and the arterio-venous bloodflows through the gills (Booth 1979; Wahlqvist and Nilsson 1980; Nilsson and Pettersson 1981).

Innervation of the Systemic Vasculature

Johnels (1956) describes an innervation of blood vessels by spinal autonomic fibres in lampetroid cyclostomes. At least some of these show catecholamine fluorescence (Falck—Hillarp technique) (Govyrin 1977) and may therefore be adrenergic. In the elasmobranch *Squalus* fluorescence histochemistry also shows the presence of adrenergic fibres in the major systemic arteries, and adrenergic drugs stimulate both alpha-adrenoceptors and beta-adrenoceptors in these vessels (Nilsson, Holmgren and Grove 1975).

In teleosts fluorescence histochemistry has shown adrenergic nerve fibres in the walls of arteries and arterioles of all dimensions in most organs. This observation, combined with pharmacological evidence for the presence of adrenergic receptors, indicates a functioning adrenergic innervation of major arteries as well as resistance vessels in individual organs. Generally the dominating effect of nerve stimulation is a vasoconstriction mediated by alpha-adrenoceptors. If this effect is blocked by a specific alpha-adrenergic antagonist, a beta-adrenergic dilation of smaller magnitude in response to nerve stimulation is often revealed (Figure 1.4; Reite 1969; Holmgren 1978; Nilsson 1983).

Opinions differ on whether the adrenergic control of systemic vascular beds is mainly neuronal or mainly humoral, exerted by circulating catecholamines (Wahlqvist and Nilsson 1977; Smith 1978).

A cholinergic innervation of the major arteries in fish has been suggested from pharmacological evidence in eel and trout (Kirby and Burnstock 1969). However, the cod coeliac artery and cod visceral and systemic vascular beds, as well as the coeliac artery from other gadid fish show low or no sensitivity to cholinergic drugs (Reite 1969; Holmgren and Nilsson 1974; Wahlqvist and Nilsson 1981; Holmgren, unpubl. obs.). Neither does nerve stimulation or transmural stimulation of the coeliac artery give any evidence of a cholinergic postganglionic innervation of this vessel (Holmgren 1978). It must be concluded that the degree of cholinergic innervation of teleost systemic arteries may vary with species. There are many examples of sensitivity to acetylcholine in different organs, but few cases where a cholinergic innervation has been shown.

Cardiovascular Reflexes

If an elasmobranch or a teleost is exposed to rapidly induced hypoxia a bradycardia immediately develops. Oxygen receptors in the gills are stimulated, afferent impulses are sent to the cardioregulator centre in the medulla, and the impulse frequency in the vagal inhibitory fibres to the heart is increased. An increase in stroke volume accompanies the decrease in heart rate. This is considered to be due to increased venous filling between beats, since no evidence has been found for an involvement of adrenergic nerves in the reflex (Satchell 1961; Butler, Taylor, Capra and Davison 1978; Jones and Randall 1978).

An increase in the blood pressure which can be induced by injection of adrenaline activates baroreceptors in teleosts as in higher vertebrates. Also in this case afferent impulses to the medulla, triggering efferent impulses in the vagus, lead to a reflex bradycardia. The baroreceptors are located in the gill region, either in various parts

Figure 1.4: Contractions of the coeliac artery of the cod (*Gadus morhua*) induced by electrical stimulation of the splanchnic nerve with 15 Hz, 1 ms pulse duration and 20 V for 30 sec. with 8 min. intervals (black dots). Addition of the alpha-adrenoceptor blocking agent phentolamine (PHENT) in a concentration of 10^{-5} M causes blockade of the contractions, and instead a weak relaxation of the artery is seen during nerve stimulation. This inhibitory effect can be blocked by beta-adrenoceptor antagonists such as propranolol and sotalol (not shown).

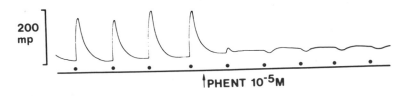

200 mp

↑PHENT 10^{-5}M

of the gills themselves or they may, in some species, be located in the pseudobranch (Mott 1951; Laurent 1967; Wood and Shelton 1980).

Autonomic Innervation of the Spleen

In elasmobranchs and teleosts the autonomic nerve fibres to the spleen run in the splanchnic nerve. In dipnoans no separate nerve to the spleen has been described, and fibres may instead run in the vagosympathetic trunk along the gut (Jenkin 1928; Abrahamsson *et al.* 1979b). The nerve fibres reaching the spleen innervate smooth muscle in the capsule and trabeculae of the spleen and the splenic vasculature. Fluorescence microscopy shows the presence of adrenergic fibres surrounding vessels and, more sparsely, in the splenic trabeculae of the elasmobranch *Squalus* and the teleost *Gadus*. In other species like *Lepisosteus* and *Huso*, the strong autofluorescence from porphyrine derivates may obscure weaker fluorescing nerve terminals, whose presence is indicated by measurement of catecholamine content in the spleen (Nilsson 1983).

Adrenergic fibres reaching the fish spleen act on alpha-adrenoceptors, producing a contraction of splenic smooth muscle. This contraction causes expulsion of erythrocytes from the spleen into the bloodstream. However, the amount of erythrocytes released is not great enough to cause a significant increase in haematocrit. A small dilatory effect produced by the beta-adrenergic drug iso-prenaline has also been described in elasmobranchs and teleosts (Nilsson 1983).

The concentration of catecholamines circulating in the blood of *Gadus, Squalus* and *Protopterus* is high enough to affect the spleen. It is even possible that in *Squalus* and *Protopterus* the circulating cate-cholamines are the major if not the only regulators of the splenic smooth muscle (Abrahamsson and Nilsson 1976; Abrahamsson *et al.* 1979b).

A cholinergic control of the fish spleen by cranial (vagal) fibres is absent. However, cholinergic receptors have been described in the spleen of some fish species. The effects mediated by these receptors are often weak and inconsistent, but in some teleosts and dipnoans stimulation of cholinergic receptors produces a contraction of the same magnitude as that produced by adrenergic stimulation (Nilsson and Grove 1974; Holmgren and Nilsson 1975; Abrahamsson *et al.* 1979b). In *Gadus* the cholinergic effect may be induced by stimulation of splanchnic nerve fibres. The effect is blocked by cholinergic antagonists, and also by treatment known to destroy *adrenergic* neurons selectively. It has been speculated that in the spleen of *Gadus* adrenaline/noradrenaline and acetylcholine are released from the same neuron (Nilsson and Grove 1974; Holmgren and Nilsson 1976; Winberg, Holmgren and Nilsson 1981; Figure 1.5).

Autonomic Control of the Swimbladder

The swimbladder is a derivative of the gut which lacks digestive and peristaltic functions and is specialised for gas storage. Humoral factors carried by the blood, such as adrenaline from chromaffin cells, may have a modulatory influence on the swimbladder functions, but the main control of gas exchange in the teleost swimbladder is exerted by autonomic nerves. The main nerve supply comes from the ramus intestinalis of the vagus. This nerve branch contains vagal fibres and fibres from the sympathetic chains, and this constitutes a 'vago-sympathetic trunk'. Fibres from the coeliac ganglion reach the swimbladder in the anterior splanchnic nerve, and there may also be other direct fibres to the swimbladder from sympathetic chain ganglia.

Physostome swimbladders are innervated by catecholamine-containing fibres, some of which reach the pneumatic duct and sphincter (Figure 1.6a). It is possible that the adrenergic fibres cause opening of the pneumatic duct sphincter. After section of the ramus intestinalis of the vagus the sphincter closes more firmly, making the fish unable to use the pneumatic duct for release of gases from, or for swallowing air into, the swimbladder. Relaxation of the smooth muscles of the pneumatic duct and sphincter may be mediated by the adrenergic fibres acting via beta-adrenoceptors, whilst the smooth muscle of the main part of the swimbladder is provided mainly with alpha-adrenoceptors mediating a contraction of the muscle (Fänge 1953, 1976). The pneumatic duct sphincter also contains striated muscle which is probably innervated by the vagus nerve.

In euphysoclists (Fänge 1953), the permeability of the swimbladder wall is regulated by the movements of the gas-impermeable mucosa-submucosa. This membrane, notably the secretory part of the swimbladder, has a very rich supply of adrenergic fibres emanating from the 'vagosympathetic trunk' (Figure 1.6b). In labrids (wrasses) and many other species the secretory part of the mucosa

Figure 1.5: Effects of surgical denervation (DEN) or 6-hydroxy-dopamine treatment (6-OH-DA) on the activity of choline acetyltransferase (ChAT) in the cod (*Gadus morhua*) spleen. 6-hydroxydopamine is known from experiments with mammals to destroy selectively the adrenergic nerve terminals, while cholinergic neurons are left intact. In the cod spleen, however, both the adrenergic and the cholinergic effect of nerve stimulation is abolished by 6-OH-DA (Holmgren and Nilsson 1976) and the activity of ChAT, which is an enzyme strictly associated with cholinergic neurons, is also reduced to a level similar to that seen after total surgical denervation. These results are compatible with the view of acetylcholine and the catecholamines stored within and released from the same neurons. The ChAT activity is expressed as nmol (1-^{14}C)-acetylcholine formed from choline and (1-^{14}C)-acetyl CoA per gram of tissue and hour. Significance levels: ** < p 0.01 (Mann—Whitney two tailed U-test).

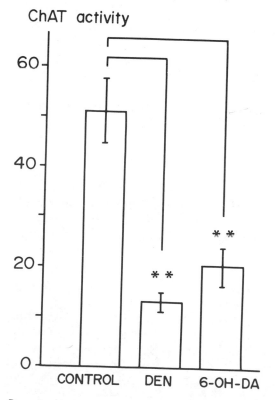

Figure 1.6: Falck—Hillarp fluorescence histochemistry of the pneumatic duct sphincter in the gut wall of the goldfish (*Carassius auratus*) (A), and a stretched preparation of the cod (*Gadus morhua*) swimbladder mucosa (B). In A varicose fluorescent fibres are seen in the musculature of the sphincter. In B fluorescent fibres can be seen following the radial smooth muscle bundles from the oval edge. In both photographs the fluorophore is due to the presence of catecholamines in the nerve fibres. Calibration bars = 100 μm.

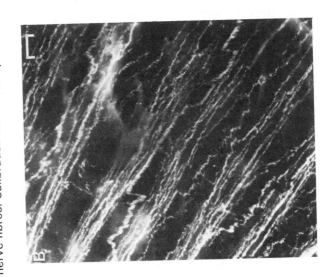

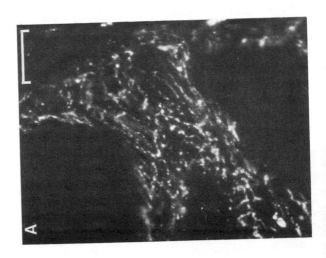

relaxes during an 'inflatory reflex', covering the inside of the swimbladder. In this state the wall of the swimbladder is relatively more impermeable to gases than during the 'deflatory reflex' when the same membrane contracts to expose the capillary network of the resorbent area to the gases of the swimbladder (Fänge 1953, 1976). Contraction of the secretory part of the mucosa is mediated by adrenergic fibres acting via alpha-adrenoceptors, whilst the resorbent part of the mucosa is instead relaxed by adrenergic fibres by a beta-adrenoceptor effect. The smooth muscle responsible for the mucosal movements forms a thin muscularis mucosa associated with the inner epithelium of the swimbladder wall (Fänge 1953, 1976).

The blood vessels of the secretory part of the euphysoclist swimbladder and the swimbladder of the eel (*Anguilla*) contract on stimulation of the swimbladder nerves, but those in the resorbent part dilate under the same circumstances. These influences seem to be due to alpha- and beta-adrenoceptor mechanisms respectively (Fänge 1953).

The gas gland receives a secretory innervation via the vagus nerve. Preganglionic fibres to the gas gland probably form synapses with postganglionic neurons in the 'gas gland ganglion', a group of nerve cells situated at the proximal pole of the counter-current vascular bundles (rete mirabilae). Bilateral vagotomy completely abolishes gas secretion (Bohr 1894), and there is strong but indirect evidence that the secretory fibres are cholinergic (Fänge 1976; Fänge and Holmgren 1982). However, no studies concerning the possibility of transmitter substances other than acetylcholine and catecholamines have so far been conducted. Possibly the swimbladder, like the gut, is provided with nerve terminals and endocrine cells containing physiologically active peptides, 5-hydroxytryptamine or other compounds.

Autonomic Control of the Melanophores

Melanophores of fish skin operate under hormonal and/or nervous control. In cyclostomes (lamprey) and elasmobranchs the melanophores probably are not innervated, and colour change is provoked by a hormone from the hypophysis (MSH). Other factors, such as melatonin from the pineal organ, may also be involved in melanophore control (Bagnara and Hadley 1973). In teleosts the melanophores are controlled by both hormones and nerves. In some species the hormonal control may predominate, in others the nervous. *Holocentrus* and *Prionotus* aggregate the pigment of erythrophores in 2—5 seconds, and *Fundulus* changes from dark to pale in less than 2 minutes. Rapid responses such as these are produced by nerve impulses (Parker 1948).

The pathways of the teleostean chromatic nerves were studied by von Frisch (1911) and followers. Results of nerve sections and electrical nerve stimulation show that the melanophores receive pigment-aggregating fibres from the sympathetic chain. The fibres reach the skin via the spinal nerves. Histochemical studies reveal an innervation of the melanophores by plexuses of varicose catecholamine-containing fibres (Falck, Münzig and Rosengren 1969).

Teleostean melanophore pigment as a rule is aggregated by adrenergic agents and dispersed by adrenergic blocking substances (Grove 1969a, and others). The adrenergic aggregating influence is mediated by alpha-adrenoceptors, but in a few teleosts beta-adrenoceptors mediate the melanophore dispersion (Fujii and Miyashita 1982). In the siluroid fish, *Parasilurus*, the cholinergic agents acetylcholine and carbachol aggregate melanophore pigment whereas dopamine disperses it (Fujii and Miyashita, 1976). It is concluded that teleostean melanophores are controlled by spinal autonomic aggregating fibres which usually are adrenergic but occasionally may be cholinergic.

In addition to aggregating nerve fibres teleostean melanophores are assumed to receive dispersing fibres, but the evidence is largely indirect, based on redarkening of faded bands after a transverse cut in the caudal fin (Parker 1948). From experiments on *Phoxinus* Grove (1969b) concluded that although darkening may be nervously controlled there is no evidence for cholinergic darkening fibres.

Fish melanophores in some ways resemble smooth muscles. The pigment granules are situated between radially-arranged microtubules, and studies with the high-voltage electron microscope show that in the nerve-controlled erythrophores of *Holocentrus* the pigment granules are connected by fine microtubular strands which contain actin and myosin. Changes of the contractile state of the microtrabecular system are probably triggered by calcium (Porter and Tucker 1981).

Innervation of the Urinary Bladder and Gonads

In the posterior trunk region, a vesicular nerve leaves each sympathetic chain and runs to the urinary bladder and the gonads (Young 1931; Nilsson 1970). This nerve contains both myelinated (preganglionic) and non-myelinated (postganglionic) fibres. An elongated vesicular nerve ganglion with mainly non-adrenergic, supposedly cholinergic, nerve cell bodies is found along its course (Young 1931; Figure 1.7).

Stimulation of the vesicular nerve causes a cholinergic contraction of both urinary bladder and gonads (Young 1936; Nilsson 1970). After treatment with atropine the contraction of the cod urinary bladder disappears and a relaxation in response to nerve stimulation is revealed. This can be blocked by treatment with drugs that deplete adrenergic nerves of transmitter, or with the beta-adrenoceptor blocking-agent propranolol. This demonstrates an inhibitory innervation by adrenergic nerves acting on beta-adrenergic receptors

Figure 1.7: A diagram showing the pattern of innervation of the cod (*Gadus morhua*) urinary bladder. In the excitatory (cholinergic) pathways (solid lines) ganglionic synapses are found in the wall of the urinary bladder (vesicular nerve ganglion). Adrenergic inhibitory fibres (broken lines) also innervate the urinary bladder and possibly the gonads. The cell bodies of the adrenergic neurons are mainly located in the sympathetic chain ganglia (not shown).

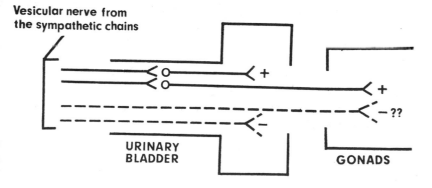

Vesicular nerve from
the sympathetic chains

URINARY BLADDER

GONADS

(Nilsson 1970). The presence of an adrenergic innervation of both the urinary bladder smooth muscle and arteries/arterioles in the cod bladder wall is further confirmed by fluorescence histochemistry (Nilsson 1973).

2 DIGESTION AND THE CONTROL OF GASTROINTESTINAL MOTILITY

S. Holmgren, D.J. Grove and D.J. Fletcher

Studies on the alimentary canal of fish have been motivated by a variety of reasons, including curiosity about how digestion and peristalsis occur, a need to predict appetite in captive or farmed fish and to allow prediction of fish production in natural waters. This review attempts to summarise our present knowledge but the interested reader will find further coverage in more extensive, multiple-author presentations such as Volume VIII of *Fish Physiology* (Hoar, Randall and Brett 1979) which deals with nutrition, digestion, bioenergetics and growth of fish.

Digestion

In their 1979 review, Fänge and Grove described the results obtained from more than 60 species of fish in which the rate of gastric emptying had been followed. Emptying rate of the stomach depends on the species and also on such factors as temperature, food type, meal size, fish size and even method and frequency of feeding. Other factors such as reproductive state of the fish, photoperiod, stock density and disease are also likely to affect digestion. Most workers sampled the stomach contents at various times after a meal so that *emptying curves* could be constructed and so allow measurements of *emptying rate* (e.g. in grams/hour). Others have not quantified the rate of emptying in this way, but have confined themselves to observing the time taken for the meal to be emptied (and present their data as *emptying times*/hours). This latter method is particularly suited to X-ray techniques (Figure 2.1). The reason for such studies depends on the idea that the fish will eat available food in amounts depending on the present stomach fullness and at intervals determined by the rate of stomach emptying.

Grove, Lozoides and Nott (1978) for example, found that appetite returns in the rainbow trout not only in close proportion to

Figure 2.1: An 18 g turbot *Scophthalmus maximus,* was fed two pellets (1 per cent b.w.) containing barium sulphate (20 per cent in Aberdeen flatfish diet). The three X-radiographs show, in sequence from top to bottom; food in the stomach only; food passing from stomach to the rectum while ca. 50 per cent of the meal is still in the stomach; empty stomach and the final residuum passing the intestine.

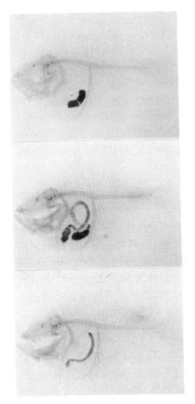

Source: Grove and Moctezuma, unpubl. obs.

the decline in stomach content, but that increased daily rates of feeding observed in fish fed on low-energey diets were achieved by more frequent meals and were accompanied by more rapid rates of gastric emptying. Jobling (1980) concluded that the overall energy content of the diet, and not deficiency of a specific nutrient, was the factor monitored by the fish. However, not all fish possess stomachs and the question arises of how appetite in these species is controlled. Grove and Crawford (1980) found that in the stomachless *Blennius pholis* the return of appetite is closely related to fullness of the anterior segment of the intestine.

On the evidence available at that time, Fänge and Grove (1979) proposed that for a given meal the rate of digestion should be primarily related to the surface area of the food bolus in the stomach. They wrote:

$$dV/dt = - K \, V^{0.67}$$

where V represents the volume of food in the stomach at a moment in time and K is the instantaneous digestion rate (which will vary with such factors as food type and temperature).

It follows from this equation that gastric emptying time (GET) will vary with the ingested meal size as:

$$GET \text{ (hours)} = K' \, V_o^{0.33}$$

and such a relationship seemed implicit in the results of Jobling, Gwyther and Grove (1977).

Since then, several workers have set out to test whether this model does apply to their data and Jobling (1981) recently reviewed these results. Jobling observes that stimulation depends on stretching the stomach circumference as food arrives so that emptying rate depends on the *square root* of the food volume:

$$dV/dt = - K \, V^{0.5} \quad \text{whence GET} = k' \, V_o^{0.5}$$

He re-analysed the published data of various authors and demonstrated that his model is in most cases an excellent description of the results. Nevertheless, if initial delays in the onset of emptying are allowed for, exponential curves do describe stomach emptying very well:

$$dV/dt = - K \, V \quad \text{whence GET} = k' \ln(V_o + 1)$$

where 'l' is a small residuum taken to represent the end-point of the exponential emptying curve (Grove and Crawford 1980) (thus if contents are expressed in mg, 1 mg is a probable limit in accuracy for most techniques). Sufficient data are now available to propose a better model for more general predictions of emptying in fish. Fänge and Grove (1979) pointed out that fish of similar habits have similar digestion times so that it should be possible to give a simple 'ecological number' to a species to indicate its digestion rate when consuming a stated food. This may be achieved as follows:

a) Gastric emptying time is made up of two factors: an initial, temperature-sensitive delay for the food type before emptying begins and an emptying phase. The delay (t_d hours) can be measured by stomach sampling.

b) Once emptying begins, until meal is emptied at time t_{end}, the stomach works at a varying rate. It is stimulated by stretch of the stomach wall by the enclosed food contents. In many fish, the stomach volume is proportional to body weight and, when filled, the stomach stretches near-isometrically in *three* dimensions. Therefore, stimulus $\propto (V/aW)^{0.33}$ and varies between 0 and 1. The secretory response of the individual fish depends on the available gastric secretory epithelium. So response (max) $\propto (a\,W)^{0.67}$. The released enzyme attacks the food at its surface, but two situations can be envisaged: if a single item is taken; or if several items fuse to make a bolus, surface area $\propto (V)^{0.67}$ where V is the total volume of food.

However, if separate items remain apart, this simple relationship does not hold. We can envisage a food item of given shape, weight V. If this were to be subdivided into m parts of the same shape, the surface area would increase by a factor of $m^{0.33}$. If n items weight V were given, surface area would increase in proportion to n. Combining these, we have in general: surface area $\propto n\,m^{0.33}\,(V)^{0.67}$ but it is useful in the later development to write s. area $\propto n^{0.33}\,m^{0.33}\,(nV)^{0.67}$.

In the 'bolus model', we can describe the gastric emptying rate as:

$$dV/dt = -k \cdot e^{bT} \cdot (V/aW)^{0.33} \cdot (W)^{0.67} \cdot (V)^{0.67}$$

$$= -k' \, e^{bT} \cdot W^{0.33} \cdot V$$

Not only is this the equation for an exponential curve if temperature and weight are held constant, but this is also an explanation of the weight exponent described by Jobling *et al.* (1977).

Gastric emptying time for the species on the stated food is therefore predicted as:

$$GET = t_d' + t_{end} = t_d' \, e^{-bT} + (1/k) \, e^{-bT} \cdot W^{-0.33} \ln(V_o + 1)$$

where t_d' is the delay at 0°C; b is the temperature coefficient (normally close to Q_{10} = 2.0 to 2.3 whence b = 0.07 to 0.08); and T is in degrees centigrade. The major variable in the equation will be the delay time for the given food (t_d') and the digestive ability of the fish (k'). Given these two values, the digestive performance of the fish under a wide variety of conditions can be predicted, provided natural prey items are known. If the species in question ingests n items weight V (which may be subdivided into m smaller items of the same shape), and these remain separate, this equation becomes:

$$GET = t_d' \, e^{-bT} + (1/k) \cdot e^{-bT} \cdot W^{0.33} \cdot n^{0.33} \cdot m^{0.33} \cdot \ln(nV_o + 1)$$

showing that subdividing each item or taking more items, or both, increases the gastric emptying rate. Increase in n also prolongs emptying times. The same two 'ecological numbers' will be required for predictions (t_d' and k).

The main point of the argument is that feeding frequency is strongly correlated with gastric or foregut emptying time. This relationship was studied in *Limanda* using demand feeders and X-radiography (Gwyther and Grove, 1981) with the result that: Interfeeding period = 1.09. (GET) — constant. In that study, the constant (2.6 hours) may represent the inability of X-ray methods to detect the last faint traces of the meal *in vivo*. Incidentally, the model also predicts that a larger fish will process a given weight of food (e.g. 1 g) faster than a smaller fish although the smaller fish processes a given relative meal (e.g. 1 per cent body weight) more quickly than bigger fish (Jobling *et al.* 1977; Flowerdew and Grove (1979).

Appetite

Despite major research efforts, our understanding of the physiology of appetite control even in mammals is still poor. The willingness of an animal to feed, the amount ingested (satiation amount) and the feeding frequency changes with a variety of factors. These include the texture and palatability of the food, its digestibility and its content of energy and nutrients. Furthermore, the previous feeding history of the individual, its size and fatness and the season of the year are also important. Most authors agree that fullness of the stomach and associated orogastric factors are major influences which induce satiation and cessation of feeding. However, the factors which determine the onset of feeding are a major subject of debate; a variety of integral mechanisms have been implicated which suppress or inhibit the willingness of the animal to accept food. Some of these factors are associated with short-term regulation of appetite (e.g. the termination of a feeding bout, the readiness to feed) whilst others act in the long term so that the body weight remains close to that typical of the species.

Fish fed infrequently, or offered food after a period of deprivation, will increase the size of the meal they take but, if the deprivation is too long, meal size is reduced when the fish first starts to feed again. In mammals, various workers have suggested that blood-borne and tissue reserves of specific nutrients are monitored by receptors in the brain and/or peripheral organs (e.g. the liver) to achieve homeostasis. The nutrient in question has been said to be glucose (glucostatic hypothesis of Mayer 1955), essential amino acids (aminostatic hypothesis of Mellinkoff *et al.* 1956), fatty acids and glycerol (lipostatic hypothesis of Kennedy 1953) or even the combined energy-donating nutrients of all types (energostatic hypothesis of Booth 1978; multifactorial model of Vanderweele and Sanderson 1976). As a consequence studies have been made of sensory stimulation, orogastric monitoring of the deglutition process, stomach and intestinal fullness, postprandial changes in circulating nutrients and hormones, the effects of nutrient infusions into selected peripheral and brain vascular beds and the effects of nerve lesions in both peripheral and central locations.

In summary, most workers accept that changes in circulation levels of glucose, fatty acids, glycerol and essential amino acids are monitored by the liver and by the brain. Similar monitoring of these nutrients occurs in the intestinal lumen. Hormones such as insulin and cholecystokinin may themselves be effective on the key centres of the brain. Early representations that discrete satiation and hunger centres lie in the ventromedial and lateral regions of the hypothalamus may be too simplistic in the light of more recent study, yet there is little doubt that the hypothalamus plays a large part in coordinating the many inputs and signals which lead to initiation or suppression of appetite. The situation in fish is in urgent need of comparable study. The simplistic models of the previous section and the direct observations of a variety of workers (e.g. Elliott 1975) do allow good predictions of appropriate feeding rates for fish in, for example, pond and laboratory conditions. As regards the physiological mechanisms which control appetite however, very little work has been done.

Fish will respond to food deprivation by an increased rate of feeding or by an increase in meal size after a short deprivation period (e.g. Tyler and Dunn 1976) and this has been interpreted as awareness of a systemic debt which must be corrected (Colgan 1973). Prolonged deprivation leads to suppression of appetite (Bilton and Robins 1973). The long-term effect could reflect the change from glycogenolysis in the liver to gluconeogenesis from other tissue stores. If the energy content of food decreases, fish have been found to quickly increase the size and/or frequency of meals (Grove *et al.* 1978; Lovell 1979) in attempts to maintain daily energy intake. Peter (1979) proposed that fish, like mammals, adjusted their body size to a 'set point' value depending on their energy stores, current size and season. The compensatory feeding increase on low-energy diets may be achieved by decreased intestinal-gastric inhibition (mediated e.g. by enterogastrone?) such that the faster emptying rate leads to an earlier return to presatiation conditions.

A consequence of ingesting and assimilating a meal is that there is a postprandial increase in metabolic rate (apparent specific dynamic action) which has led several authors to include this factor in appetite control (Colgan 1973; Vahl 1979). The effect is mainly due to deamination of amino acids and their transformation into new metabolites such as glycogen or fat. The return of appetite may be delayed if postprandial circulating levels of amino acids cannot be metabolised readily, for example if the limited oxygen supply must simultaneously meet the need for basal metabolism and increased activity. Such control could be envisaged by reducing blood supply to the gut, slowing digestion and hence prolonging the signals which maintain satiation.

Research into the effects of circulating nutrients on fish appetite are few. Kuzimina (1966) showed that infusions of glucose or essential amino acids depressed appetite in the carp, but that non-essential amino acids or saline infusions did not. However, attempts to correlate blood nutrient concentrations with appetite have so far met with mixed success. Bellamy (1968) presented evidence that shows that the lowest plasma glucose levels in *Rooseveltia natterei* precede the period of greatest intake whereas Magnusson (1969) found that *Katsuwonus* could continue to feed when blood glucose levels are rising. Peter, Monckton and McKeown (1976) could not detect the presence of glucoreceptors in the brain of *Carassius* although it is possible that such receptors could be located in peripheral organs such as the liver. Lovell (1979) found that *neither* stomach fullness nor serum levels of nutrients appeared to be correlated with the appetite of *Ictalurus*. The studies of Page and Andrews (1973) and of Lovell (1979) indicate the difficulties that beset researchers in this field. If dietary energy is raised by increasing the lipid or carbohydrate component of the diet (to raise the digestible energy/protein ratio) food intake decreases. This ensures that calorie intake remains constant. However, food intake is not necessarily controlled simply by the calorific content of the diet. When isocalorific diets are prepared, increase in protein content from 15 to 45 per cent had little detectable effect on feeding but, above this level, intake declines. This indicates some deleterious effect of high dietary protein levels.

Many species of fish show seasonal mobilisation or repletion of body energy reserves, as well as cyclical changes in both gonads and true body growth. Pickford and Atz (1957) report that bovine growth hormone not only stimulated growth of hypophysectomised *Fundulus* but also stimulated their appetite. The suggestion has also been made that such hormones as thyroxine and prolactin (which are implicated in growth control and lipid storage) may themselves control appetite, at least indirectly. Purdom (1979) recounts that many fish cease to feed during the breeding season but it is not known whether gonadotrophins and sex steroids have direct effects on appetite.

Studies are urgently required on postprandial levels of respiration, circulating nutrients, circulating hormones as well as on digestion rates for single species fed a variety of diets, using fish of stated size, sex and reproductive state under clearly-defined husbandry conditions. Only when this level of experimental design is achieved will reasonable progress toward understanding fish appetite be achieved.

Despite these constraints, several workers have taken the step to attempt estimations of fish appetite in natural waters. The techniques adopted usually include measurements of stomach contents of fish taken sequentially through the day. In addition, information of digestion rates or stomach emptying rates (obtained in the laboratory or in the field) is used to convert stomach contents to feeding rates. The papers of Elliott and Persson (1978) and of Jobling (1981) are of interest to workers in ecology and fisheries.

The Control of Gastrointestinal Motility

The detailed structure of the alimentary canal of fish was reviewed recently (Kapoor, Smit and Verighina 1975; Fänge and Grove 1979) and the general pattern emerged that the major divisions of this organ system, and its histology, conform to the patterns seen in other vertebrates. The digestive tube consists of an inner mucosa, supporting submucosa, smooth muscle coat and outer serous layer. Major adaptations occur in different fish according to their feeding habits and greatest variation is seen in the structure of the mouth, pharynx, stomach (when present), the length of the intestine and in the enzyme complement associated with mucosal- and digestive gland-cells. As in other vertebrates a large population of nerve cells occurs in the gut wall, mainly lying in the myenteric plexus (Auerbach's plexus) between the inner circular and outer longitudinal smooth muscle coats. This enteric nervous system coordinates peristalsis, mixing movements, blood flow and other complex activities of the canal. In addition, mucosal cells are present which produce hormones (e.g. gastrin, secretin, cholecystokinin) to affect both nearby and distant segments of the gut, whilst nerves and hormones from other parts of the body can modify gastrointestinal function when such need arises.

Extrinsic Nerves Which Affect Gastrointestinal Function

Vagus Nerve.   In cyclostomes, elasmobranchs and teleosts the vagus nerve sends branches to the gut. However, in jawless fish little effect of electrical stimulation has been observed, whereas this nerve exerts profound effects on the oesophagus and stomach in jawed fish. Campbell (1970b) took issue with earlier reviewers who had maintained that stimulation of the vagus branch induces contractions of the stomach (in those fish which possess this organ). He pointed out that such stimulation could excite nerves coming from the adjacent sympathetic system and which accompany the vagus to the periphery. Furthermore, he pointed out that 'rebound contractions' of smooth muscles which occur at the offset of electrical stimulation of *inhibitory* nerves had been demonstrated in mammals. The inhibitory nerves in tetrapods release an unknown transmitter (non-adrenergic/non-cholinergic; NANC) and he postulated such nerves to be also present in the fish vagus. In 1975 the same author demonstrated such inhibitory nerves in the vagus of *Scyliorhinus*, *Salmo* and *Conger*. In the dogfish, this inhibition is likely to involve only the inner circular smooth muscle layer since Young (1980a) could not detect an effect on longitudinal muscle of *Scyliorhinus* or *Raia.* In the rainbow trout, Holmgren and Nilsson (1981) have also detected a small vagal excitatory component to the stomach which, because it is blocked by atropine, is presumably mediated by acetylcholine. More advanced teleosts appear to differ, since intracranial stimulation of the vagus roots in *Pleuronectes* (Stevenson and Grove 1977) demonstrates both NANC inhibitory fibres and, in addition, cholinergic excitatory fibres passing to both stomach and oesophagus. Grove and Campell (1979b) found vagal excitatory fibres in the advanced teleost *Platycephalus* but were not able to detect inhibitory nerves following this pathway. Nilsson and Fänge (1969) also found only excitatory effects of the peripheral vagus tract on the stomach of *Gadus.* A pattern appears which suggests that primitive fish have a predominantly inhibitory vagal control over the alimentary canal, which may then have a superimposed or dominant excitatory component in the more recent bony fish. The recent report of such a dual vagal system in the advanced fish *Lophius* (Young 1980b) reinforces this idea. Campbell (1970b) pointed out that the vagus had been reported to extend its influence to the intestine of stomachless fish. However, this is true of certan Cyprinidae (e.g. *Tinca*) but is not true for the stomachless flatfish *Rhombosolea* and *Ammotretis* (Grove and Campbell 1979a). These latter fish appear to belong to a sequence of species in the Pleuronectidae in which there is a progressive reduction in the size of the stomach.

Fänge and Grove (1979) reviewed the work on the excitatory nerves of the fish vagus, which are cholinergic for the most part, but progress in understanding the inhibitory neurotransmitters has been slow. Holmgren and Nilsson (1981) have shown that the inhibitory transmitter from *Salmo gairdneri* vagal stimulation can be detected on muscle strips by superfusion and its effects are not prevented by adrenoceptor (phentolamine) or muscarinic receptor (atropine) blocking agents. However, the vagus nerve fibres are probably preganglionic, releasing acetylcholine in the stomach wall to activate intrinsic neurons since the ganglion blocker chlorisondamine abolishes the effect.

Sympathetic Nerves. In the elasmobranch *Scyliorhinus,* Young (1980a) reports that anterior splanchnic nerve stimulation either inhibits (low frequency) or excites (high frequency) the stomach and that these effects are not impaired by ganglionic-blocking drugs (nicotine, hexamethonium). The excitatory component to the pylorus may be due to release of adrenaline since phentolamine blocks the nerve effect and adrenaline in itself is excitatory in high doses. The cardiac part of the stomach is insensitive to adrenaline, and the nerve-mediated stimulation is unaffected by adrenergic neuron-blocking agents. In another elasmobranch, *Squalus*, both longitudinal and circular preparations of cardiac or pyloric stomach wall are sensitive to adrenaline (Nilsson and Holmgren, in prep.) and abundant adrenergic tracts have been observed in the myenteric plexus (Holmgren and Nilsson, in prep.).

The type of neurons mediating inhibition are not yet known since acetylcholine and adenosinetriphosphate are excitatory. Interestingly, 5-hydroxytryptamine has weak inhibitory effects at low doses but is excitatory at higher doses; it merits further study. Radioimmunoassay techniques have shown the presence of vasoactive intestinal polypeptide (VIP) in extracts from all parts of the gut of *Scyliorhinus canicula* (Foucherau-Peron, Laburthe, Besson, Rosseline and Le Gal 1980). Immunohistochemical studies have revealed VIP-like immunoreactivity in nerves of the myenteric plexus and along blood vessels in the stomach of *Squalus* (Holmgren and Nilsson, in prep.). VIP has been proposed as a likely inhibitory neurotransmitter in the gut of mammals (Furnoss and Costa 1980). The dual nature of sympathetic control of the elasmobranch stomach may reflect the minor role played by the vagus in these fish.

Very little information exists on the action of the sympathetic system on the intestine of these fish. Electrical stimulation of the posterior splanchnic nerves was said to be excitatory by Young (1933) but the possibility remains that rebound contractions contribute to this response. Acetycholine and 5-hydroxytryptamine are excitatory whereas adrenaline was reported to have both excitatory and inhibitory effects.

In the teleosts, the sympathetic nerve branches after leaving the coeliac ganglion and supplies the stomach (if present) and the intestine. Campbell and Burnstock (1968) concluded that on the available evidence this tract carries both excitatory and inhibitory influences. These were thought to be mediated respectively by cholinergic and adrenergic fibres. In *Salmo* (Campbell and Gannon 1976) and *Platycephalus* (Grove and Campbell 1979b) the nerve is excitatory and is blocked by phentolamine, suggesting that adrenaline is released to act on alpha-adrenoceptors. In *Pleuronectes* the sympathetic nerve tract does carry both excitatory (cholinergic) and inhibitory (adrenergic) fibres to the stomach (Stevenson and Grove 1978). The former are blocked by atropine and the latter by the beta$_2$-blocker, butoxamine, and by reserpine. Clearly the effects of adrenaline and of sympathetic nerves differ in the stomachs of different species. Adrenaline and noradrenaline excite the stomachs of *Anguilla* and *Gadus* but inhibit those of *Lophius* and *Uranoscopus* (Nilsson and Fänge 1969; Young 1980a,b). At least the pathways of the fibres containing catecholamines are more predictable and fibres passing from the sympathetic system to the stomach wall (myenteric plexus and mainly circular muscle) have been demonstrated by fluorescence histochemistry in many species.

The intestine of teleosts typically receives only a sympathetic nerve supply, although in Cyprinidae the vagus nerve is shown to innervate the striated and smooth muscle coats and causes excitation and inhibition respectively. Weak vagal influences have been reported on the most anterior segments of the intestine in *Ammotretis, Rhombosolea* and *Lophius*. The major supply of nerves to the teleost intestine comes from the sympathetic system. In *Rhombosolea* and *Ammotretis* paravascular stimulation of this tract caused inhibition (or rebound contractions) at frequencies as low as 0.5 Hz and which were readily blocked by propranolol; apparently, as in other fish (Fänge and Grove 1979), beta-adrenoceptors are involved.

The role of extrinsic nerves to the fish gut serves to enhance or decrease its intrinsic activity. Sympathetic discharge, or the arrival of circulating adrenaline, primarily abolishes peristalsis and myogenic activity and also induces vasoconstriction. The vagal gastric inhibitory nerves may be the efferent limb of the reflex which relaxes the stomach when a meal is ingested. Vagal excitatory fibres may control tone, or even emesis if noxious food is ingested.

The Myenteric Plexus

Fänge and Grove (1979) indicated that the vast majority of the neurons of the gut wall are reported to lie in the myenteric plexus between the two sheaths of smooth muscle. These nerve cells not only receive inputs from extrinsic nerves but also innervate muscle and gland cells and may innervate each other. The plexus is involved in sensory as well as other functions, although there is as yet very little evidence about integration within the network; it is not known which cells act as sensory nerves to signal gastric fullness to the brain nor even if the more localised peristaltic movements involve the same 'preparatory' and 'expulsion' phases that have been long described in mammals.

In their study of gastric peristalsis in *Pleuronectes*, Stevenson and Grove (1977) found that distension of the isolated stomach elicits rings of contraction which move aborally; when treated with tetrodotoxin to abolish nerve activity, non-propagating contractions frequently appear. They proposed that these were neurogenic and myogenic respectively. Catecholamines suppress both types of activity by way of beta-adrenoceptors. Acetylcholine and 5-hydroxytryptamine excite the muscle, even after tetrodotoxin, indicating that they act on or close to the smooth muscles themselves. Mecamylamine, the ganglionic blocker, does not always abolish peristalsis and atropine rarely affects peristalsis once it is established. Adrenergic blocking agents were without effect. In the rainbow trout, adenosine triphosphate contracts gastric longitudinal muscle but, on the circular muscle coat, the same agent has an inhibitory effect in low doses but this changes to contraction at higher concentrations (Holmgren 1982). Clearly, a wider range of drugs and of physiological experiments are required to unravel the complex interactions among enteric neurons which control gastric motility.

A similar problem occurs in studies on the fish intestine. In *Rhombosolea* and *Ammotretis*, transmural stimulation of the intestine demonstrates that, in addition to the known extrinsic nerve components, other nerve types are present. Atropine-sensitive excitatory neurons respond to stimulation frequencies above 5 Hz but at lower frequencies NANC-type responses were detected. The NANC excitatory nervous effect on the longitudinal muscle could be mimicked by 5-hydroxytryptamine and by ATP but the inhibitory effect on the circular muscle could not be further characterised. In *Pleuronectes* intestine a similar inhibitory effect was mimicked by ATP and here again qualitative differences between closely-related species can be identified.

In view of the paucity of information about the types of neuron present in the fish enteric plexus, a number of workers have undertaken appropriate histological and histochemical studies as a necessary prerequisite for physiological studies.

Earlier workers demonstrated adrenergic neurons of extrinsic origin in the alimentary canal of *Salmo*, *Anguilla*, *Carrassius* and *Tinca* (Baumgarten 1967; Read and Burnstock 1968a, b, 1969; Saito 1973). Mucosal enterochromaffin cells with the typical fluorescence of 5-hydroxytryptamine have been described in flatfish (Grove and Campbell 1979a). In addition, neurons containing 5-hydroxytryptamine have been described in *Lampetra* (Baumgarten, Bjorklünd, Nobin and Rosengren 1973), *Pleuronectes* (Watson 1979) and *Platycephalus* (Anderson, pers. comm.).

Watson (1981) went on to describe the ultrastructure of nerve terminals in the intestinal wall of *Pleuronectes* and *Myoxocephalus*. He reported, as did Wong and Tan (1978) for the stomach of *Chelmon*, the presence of only two types of nerve terminals, one with mainly small (40-60 μm) agranular vesicles (present in the enteric plexuses), the other with predominantly 50-150 μm granular vesicles which could be stained for amines (present in both enteric plexuses and the circular muscle layer). The two types of terminals were suggested to be cholinergic and adrenergic respectively. In the trout stomach (Santer and Holmgren, in prep.; Figure 2.2) and intestine (Ezeasor 1979) three types of nerve terminals are found, namely (using the terminology of Campbell and Gibbins 1979) c-type terminals containing mainly small (40-60 μm) agranular vesicles, a-type terminals containing mainly small (40-60 μm) granular vesicles and p-type terminals containing large (90-160 μm) granular vesicles. These types of terminals have by many authors been regarded as cholinergic, adrenergic/tryptaminergic and peptidergic/purinergic respectively, although the correlation between ultrastructure and physiological-pharmacological properties may not be all that clear in many cases (Gibbins 1982).

The complexity of the enteric system in higher vertebrates is indicated in the review by Furness and Costa (1980) in which more than 10 different types of enteric neurons are now recognised among the large population (numbering in excess of 10^7) in the gut wall. For a similar understanding of this system in fish, there is need for studies involving electrical, pharmacological, histochemical and ultrastructual techniques. In addition to examining extrinsic excitatory and inhibitory nerves, similar intrinsic neurons and their interneurons, the sensory elements of the peristaltic reflex, supplies to blood vessels and glands all have yet to be identified. Some progress is being made to categorise the newer types of NANC nerves which occur in the fish gut.

Several workers have investigated the possibility that such neurons might contain polypeptides of the types found in the mammalian alimentary tract. One of the most interesting of these is vasoactive intestinal polypeptide (VIP), since it has, on good evidence, been

Figure 2.2: Nerve terminals in the myenteric plexus of the rainbow trout *Salmo gairdneri*, showing (top) the presence of predominantly small clear vesicles (c-type terminal) and (bottom) predominantly large granular vesicles (p-type terminal). Magnification in both figures × 36 000.

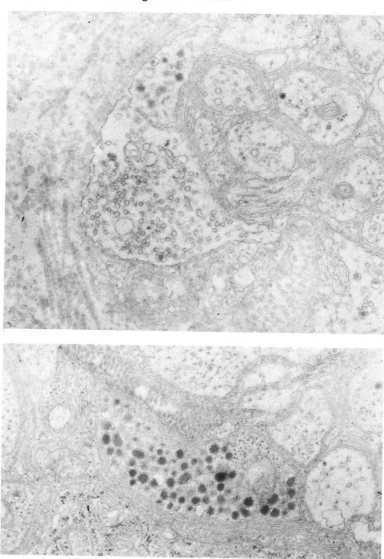

Courtesy of R. Santer and S. Holmgren

proposed to be an inhibitory NANC transmitter in the mammalian gut (see Furness and Costa 1980), and it is therefore a possibility that VIP might be involved also in the NANC-inhibitory transmission that is evoked by vagal stimulation in certain teleosts. VIP-like immunoreactivity (IR) has been found in nerves of the enteric plexuses and muscle layers of the gut of several fish species, e.g. the elasmobranch *Squalus*, the ganoid fish *Lepisosteus* as well as in a number of other teleosts (Figure 2.3A), (Langer *et al.* 1979; Van Noorden and Patent 1980; Holmgren *et al.* 1982; Holmgren and Nilsson, in prep.).

A problem when doing immunochemical studies in fish is that most antibodies used for tracing the polypeptide in question are raised against a polypeptide isolated from mammalian (and in some cases amphibian) sources. However, in the case of VIP, studies made on VIP extracts from fish gut show immunoreactive properties similar to those of mammalian VIP, indicating a close similarity in the 'active structure' of mammalian and fish VIP (Foucherau-Peron *et al.* 1980). Addition of porcine VIP in some cases produces an inhibition of spontaneous or induced activity of trout stomach strips, while in other cases an excitatory effect may be obtained (indicating a similar dual effect of VIP as described for mammals; Furness and Costa 1980; Holmgren 1982).

Another peptide with inhibitory effect, e.g. on the mammalian colon (Kitabgi and Vincent 1981), is neurotensin. A dense plexus of fibres showing neurotensin-like IR is found in *Lepisosteus* and in some other teleosts (Langer *et al.* 1979; Holmgren and Nilsson, in prep.). Preliminary studies on both longitudinal and circular muscle strips of the rainbow trout stomach indicate that neurotensin may have an inhibitory effect on the activity of the fish stomach (Holmgren, in prep.).

Somatostatin has been reported to be without direct effects on mammalian gut smooth muscle; it causes inhibition both through excitation of inhibitory enteric neurons and inhibition of excitatory neurons. In fish, somatostatin-like IR in nerve fibres of the gut has so far been shown only in *Squalus* (Holmgren and Nilsson, in prep.).

Amongst the many peptides showing direct or indirect excitatory effects on the mammalian gut smooth muscles are substance P, bombesin and enkephalin. Substance P-like, (Figure 2.3B) bombesin-like and enkephalin-like IR in nerves of the gut have also been reported in some fish species (Langer *et al.* 1979; Van Noorden and Patent 1980; Holmgren *et al.* 1982; Holmgren and Nilsson, in

Figure 2.3: (A) VIP-like immunoreactivity in nerves of the myenteric plexus of the intestine of rainbow trout, *Salmo gairdneri.* (Calibration bar = 50 µm) (B) Substance P-like immunoreactivity in nerves of the myenteric plexus of the stomach of the rainbow trout. (Calibration bar = 25 µm) (C) Substance P-like immunoreactivity in a mucosal cell in the stomach of the rainbow trout. (Calibration bar = 25 µm).

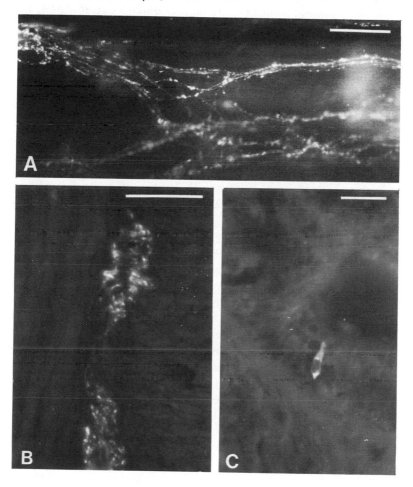

All figures by courtesy of S. Holmgren and C. Vaillant

prep.) and substance P is shown to have excitatory effects on the trout stomach (Holmgren 1982).

It should be pointed out that the results from immunohisto-chemical studies reported above indicate great variations in occur-rence and distribution of the different peptides between different fish species. This may show a true difference in innervation by the different peptidergic nerves of the fish gut. However, a negative result in an immunohistochemical study may only mean that the antibody used (raised against a mammalian peptide) fails to react with the fish peptide in the studied species. On the other hand, a positive result can only be trusted if careful specificity tests have been performed. Nevertheless, the part played by peptide neurotransmitters is likely to be a challenging and fruitful research topic in studies of fish gastro-intestinal motility.

Endocrine and Paracrine Cells

Cells in the mucosa have been found to contain catecholamines, 5-HT or polypeptides. These may directly or indirectly affect the chemical processes of digestion, but may also affect gastrointestinal motility. The cells are either endocrine, releasing their contents into the lumen of the stomach or to the blood passing through the mucosa, or they are paracrine with processes reaching nearby cells (Larsson 1980). Among the peptides that have been located in mucosal cells of the fish gut are bombesin, substance P (Figure 2.3C), gastrin, chole-cystokinin (CCK), somatostatin, enkephalin and neurotensin (Langer *et al.* 1979; Van Noorden and Patent 1980; Holmgren *et al.* 1982; Larsson and Rehfeld 1978; Holmqvist, Dockray, Rosenquist and Walsh 1979; Noillac-Depeyre and Hollande 1981; Rombout 1977; Dubois, Billard, Breton and Peter 1979; Reinecke, Carraway, Faulkner, Feurle and Forssmann 1980). As yet, very little is known about the role of these newly-discovered substances in the functional coordination of the gut of lower vertebrates.

3 CONTROL OF RESPIRATION AND CIRCULATION

P.J. Butler and J.D. Metcalfe

Although the circulatory system of vertebrates serves many different functions, the primary role of the respiratory system is to maintain an adequate supply of oxygen at the gas exchange surface and to remove carbon dioxide. Thus, when considering these two systems together it is reasonable to do so in terms of gas exchange and the supply of and demand for oxygen, for example during a reduction in oxygen availability (hypoxia) and exercise. After considering the physical and morphological factors affecting gas exchange, the control of the respiratory and circulatory systems is discussed in terms of these factors and in terms of supplying blood to the metabolising tissues.

Factors Affecting Gas Exchange

Gas exchange occurs by diffusion and the factors affecting the diffusion of a gas across a permeable membrane are given by a modification of Fick's first law of diffusion:

$$\dot{M} = KA\Delta P/x \qquad (3\text{-}1)$$

where $\dot{M}$ = amount of gas diffusing per unit time (mmol min^{-1})

K = Krogh's diffusion constant (nmol s^{-1} kPa^{-1})

A = area of membrane (mm^2)

ΔP = difference in partial pressure of gas on each side of the membrane (kPa)

x = thickness of membrane (μm)

It is important to note that, although the diffusion of the gas is measured in terms of the *amount*, the driving force is the difference in *partial pressure*. When two different fluids (e.g. water and blood) are present on either side of the permeable membrane they may have different capacitance coefficients (β) for the gas, so that the concentration of the gas at a similar partial pressure would be

41

different in the two fluids. (NB capacitance coefficient is the increase in concentration per unit increase in partial pressure.) This would give rise to the situation where gas (oxygen) is moving down a pressure gradient (from water to blood) but against a concentration gradient (Dejours 1981). This difference in β between blood and water (it being higher in blood because of the presence of haemoglobin) has other consequences. The driving pressure (ΔP) for gas diffusion is maintained as large as possible by the movement (convection) of blood and water either side of the gas exchange surface of the gill. They move in opposite directions (countercurrent) across the secondary lamellae, which allows very effective gas exchange (Piiper and Scheid 1977). This is optimum when

$$(\dot{V}_w \times \beta_w)/(\dot{V}_b \times \beta_b) = 1 \qquad (3\text{-}2)$$

where $\dot{V}_w$ = volume of water flow (ventilation volume, ml min^{-1})
  $\dot{V}_b$ = volume of blood flow (cardiac output, ml min^{-1})

As $\beta_w < \beta_b$, $\dot{V}_w/\dot{V}_b$ is 10-20 in most fish (except in the ice fish, which have no haemoglobin). In other words, a greater volume of water must pass over the gills than blood perfuse them for this condition to exist. Maintaining an adequate relationship between gross ventilation and perfusion is not the only factor affecting gas exchange. The ability of the gills to transfer gas can be determined by dividing the rate of gas exchange ($\dot{M}$) by ΔP

$$\text{giving} = \dot{M}/\Delta P \qquad (3\text{-}3)$$

and this is known as the transfer factor, T, (Randall, Holeton and Stevens 1967). It can be seen from equation (3-1) that

$$\dot{M}/\Delta P = KA/x \qquad (3\text{-}4)$$

$\therefore$

$$T = KA/x \qquad (3\text{-}5)$$

Any change in the transfer factor, therefore, indicates a change in one or more of the diffusion constant, the surface area of the gills or the thickness of the barrier separating the blood from the water.

The Gills and Respiratory System

Each gill arch of a fish has two rows of filaments, which in elasmobranchs are joined along almost their whole length by a septum which forms an external flap (Randall 1982), but which are separate for most of their length in teleosts (Figure 3.1). Each gill filament bears a series of projections (secondary lamellae) on its dorsal and ventral surfaces and it is across these secondary lamellae that gas exchange occurs. Obviously the secondary lamellae greatly increase the surface area of the gills in comparison with the volume that they occupy. However, the surface area and the mean distance between blood and water (diffusion distance) do vary between fishes. For example in the bottom-dwelling, scavenging dogfish (*Scyliorhinus canicula*), gill area is $210 \text{ mm}^2 \text{ g}^{-1}$ and diffusion distance is 11.3 μm, whereas in the very active skipjack tuna (*Katsuwonus pelamis*) the values are 1350 $\text{mm}^2 \text{ g}^{-1}$ and 0.6 μm respectively (Hughes and Morgan 1973). As might be expected, T_{O_2} is an order of magnitude greater in the continuously swimming tuna (Butler 1976).

Water flows from the buccal cavity (orobranchial cavity in elasmobranchs) through the sieve produced by the secondary lamellae and into the opercular cavity (parabranchial cavities in elasmobranchs). Although the details of the respiratory system vary from fish to fish, generally, water is forced across the gills by the active closure of the mouth and then sucked across by the active opening of the opercula in teleosts or by the passive expansion of the parabranchial cavities in elasmobranchs. However, there is no anatomical basis for there being two separate pumps. Because of interaction between skeletal elements, at least in teleosts, any one of several muscles may influence the volume of both the buccal and opercular cavities (Shelton 1970). The respiratory muscles are innervated by motor nerves from the central nervous system and rely upon activity in these nerves to elicit their contraction. Changes in the depth of respiration are normally related to the number of active muscles.

The Central Respiratory Neurons

Transection experiments have demonstrated that rhythmic respiration in fishes is still present when the medulla oblongata is isolated from the rest of the brain and spinal cord (Shelton 1970), i.e. all motor information is in the cranial nerves. There may be some changes in respiratory pattern, indicating that other regions of the brain may be involved in the control of ventilation (Ballintijn 1982).

Figure 3.1: Diagrams showing the general anatomy of gills in fish.
(a) Transverse section through a gill arch of a teleost showing
blood vessels in the filaments. Arrows indicate direction of blood
flow. (b) Three-dimensional arrangement of secondary lamellae
on part of a filament which is sectioned through a secondary
lamella to show arrangement of blood channels (in black). (Arrows
indicate direction of water flow.) (c) Transverse section through a
secondary lamella to illustrate arrangement of blood channels.
Note the marginal channel at the tip and the basal channels which
may be well within the filament. ABA: afferent branchial artery,
EBA: efferent branchial artery, AFA: afferent filament artery, EFA:
efferent filament artery, ALA: afferent lamellar artery, ELA:
efferent lamellar artery, SL: secondary lamellae.

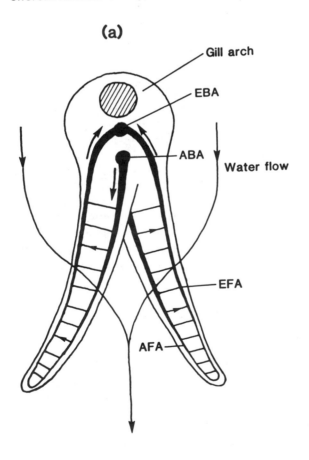

(b)

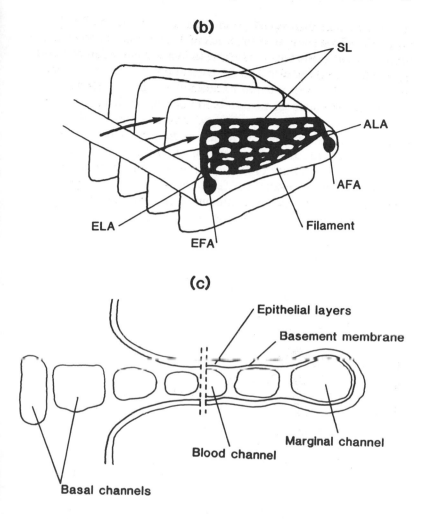

SL

ALA

AFA

Filament

ELA

EFA

(c)

Epithelial layers

Basement membrane

Marginal channel

Blood channel

Basal channels

The recording of electrical activity from the brain of teleosts, by several authors, has revealed that respiratory neurons (i.e. those that fire in phase with some part of the respiratory cycle) exist in two longitudinal strips which run the whole length of the medulla on each side of the midline (Figure 3.2). Each strip consists of the trigeminal (V), facial (VII), glossopharyngeal (IX) and vagal (X) motor nuclei, with the descending trigeminal nucleus and reticular formation alongside these nuclei. Extramedullary neurons with a respiratory rhythm have also been identified in unanaesthetised carp, in the mesencephalon, diencephalon and cerebellum (Ballintijn, Luiten and Jüch 1979). These are concerned with processing proprioceptive input (see later) and with correcting the visual image for respiratory displacements of the eye. The concentration of respiratory neurons varies, particularly in relation to anaesthesia and oxygenation, and the activity of individual medullary motor neurons may be related to the intensity of respiratory activity. There is also evidence that activity may shift among the motor neurons within the pool of a particular muscle. Neurons are not clustered according to their phase relationship with respiratory activity. So, the basic generator of the respiratory rhythm in fishes resides in the medulla, most likely in the reticular formation, but what influences the activity of this generator and the motor output?

Sensory Input to the Central Respiratory Neurons

Some central respiratory neurons continue their activity after the fish has been paralysed and must be part of the rhythm generator or be dominated by it, whereas others cease their activity during paralysis (Ballintijn 1972): the latter are therefore involved in processing mechanical information from the respiratory system. Medullary neurons responding to stimulation of length or tension receptors in muscles have been identified in the carp and are, no doubt, concerned with 'load matching'. Some of these neurons behave as if they are elements of a simple proprioceptive control arc involving muscles of one antagonistic pair and acting in response to a brief stimulus, within one respiratory cycle. Other of these neurons appear to receive information from muscles that are not antagonistic pairs and they may affect the rhythm generator over several respiratory cycles via a higher (i.e. extramedullary) level of integration (Ballintijn and Roberts 1976). In the case of the 'simple' reflex, sensory input from proprioceptors (via V and VII nerves) projects to the descending trigeminal nucleus and then via ipsi- or contralateral interneurons to the trigeminal and facial motor neurons (Figure 3.2).

The gill arches of teleosts and elasmobranchs possess numerous mechanoreceptors both on the filaments and the rakers. They are innervated by the glossopharyngeal (IX) nerve on the first gill arch and by the vagus (X) nerve on the other arches. Their continued stimulation has an inhibitory effect on respiration in the dogfish (*Squalus lebruni*) while in carp, brief electrical stimulation of an epibranchial vagal ganglion abruptly terminates abduction (expansion or 'inspiration'). Vagal information may, therefore,

Figure 3.2: Diagram showing respiratory neuron population (right) and important connections (left) in the posterior part of the brain of a teleost. TRN: tegmental respiratory neurons, RF: reticular formation, Vmn: trigeminal (V) motor nucleus, D.Vn: descending trigeminal nucleus, VIImn: facial (VII) motor nucleus, IXmn: glossopharyngeal (IX) motor nucleus, I.VIIn: intermediate facial nucleus, Xmn: vagal (X) motor nucleus, P: muscle proprioceptive input, X: vagal sensory input.

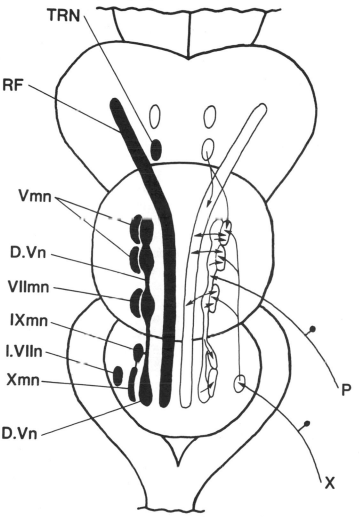

After Ballintijn 1982

contribute to the switching of respiratory phases (as in mammals). It is also possible that vagal input is important in the timing of respiration, since continual stimulation of one vagal ganglion at certain frequencies can shift respiratory frequency to a limited extent (Ballintijn 1982). The vagal input projects, via the intermediate facial nucleus, directly onto the V and VII motor nuclei (Figure 3.2). Thus, mechanoreceptors from the major respiratory muscles or the gill arches are important in controlling rate and depth of respiration in terms of load on the muscles and distortion of the respiratory apparatus, thus presumably ensuring minimum energy expenditure for a given ventilation volume. What controls respiration in response to changes in supply of or demand for oxygen? Unfortunately the answer is not so clear.

Ventilation in fishes changes in response to changes in oxygen tension and on the basis of homology with the mammalian system, it might be expected that low oxygen tension (PO_2) is monitored by chemoreceptors on the gill arches, particularly perhaps the first gill arch or the pseudobranch, innervated by cranial nerve IX (i.e. corresponding to the carotid body in mammals). Although it is possible to record an increase in afferent information from nerves innervating the isolated, saline perfused pseudobranch in response to a decrease in PO_2, bilateral denervation of this structure has no significant effect on the ventilatory response of the trout to either an increase in PO_2 (hyperoxia) or to a decrease in PO_2 (hypoxia). Even after bilateral section of cranial nerves IX and X, the ventilatory response to hypoxia is not abolished in the sea raven (*Hemitripterus americanus*). Receptors innervated by cranial nerves V and VI may be involved in the respiratory response to hypoxia, but the most likely location is the brain itself (Bamford 1974). The receptors may respond to oxygen concentration or rate of delivery of oxygen, rather than to PO_2 alone (Randall 1982). The increase in ventilation in response to a moderate increase in CO_2 tension in the water (hypercapnia) can also be explained on the basis of a reduction in blood oxygen content (Root effect), although at high levels of CO_2, there appears to be a direct effect of P_{CO_2} or pH on ventilation (Randall 1982).

It is often assumed that, as in most birds and mammals, respiratory activity in fishes is continuous. This is certainly not the case for the Port Jackson shark or the carp (Capra 1976; Lomholt and Johansen 1979). When undisturbed in well-aerated water, both of these fishes exhibit respiratory pauses (apnoea) and, in the shark, heart rate is reduced during these pauses. A reduction in environmental oxygen tension leads to more continuous respiratory activity. Such a response is, perhaps, not too surprising as far as carp are concerned. They live in ponds or lakes where they experience great diurnal and seasonal changes in environmental PO_2. They are extremely tolerant to hypoxia, and the set point of their control system may be lower than that of a trout which inhabits well-aerated water. Well-aerated water may be physiologically equivalent to hyperoxia in the carp. Certainly, hyperoxia (i.e. $PO_2 > 21$ kPa) causes periods of apnoea in the trout and the dogfish (*Scyliorhinus stellaris*).

At the onset of exercise respiratory frequency changes immediately, which indicates either that output from a common region in the brain may activate respiratory and locomotory motor neurons, or that the central respiratory neurons are excited by feedback from receptors stimulated by body movement (Jones and Randall 1978). In some fishes, respiration and locomotion may be synchronised and reticular motor neurons have been described in the dogfish (*Squalus acanthias*), which may be part of a proprioceptive reflex, linking the medulla and the spinal cord, permitting the motor patterns of one to be transferred to the other (Satchell 1968). The advantage of synchrony between respiration and locomotion is that opening of the mouth during the forward propulsive phase would augment normal respiratory activity.

The Circulatory System

The blood of fish is pumped by the heart firstly to the gills where gaseous and osmoregulatory exchange occur, and then flows on to the systemic circulation before returning to the heart. This description of the so called 'single circulation' of fish is to some extent rather an oversimplification, and is a little misleading. It does however provide a useful framework from which to describe its control. A more complete basic scheme of the circulation in fish is illustrated in Figure 3.3. As can be seen, the major differences between this and the simple description given above are that not all cardiac output entering the gills passes directly onto the systemic circulation, and not all gas exchange occurs at the gills. Even so, it should be evident that the sites at which blood flow might be controlled are the heart, within the gills and between the various vascular beds of the systemic circulation. These areas will be considered in turn.

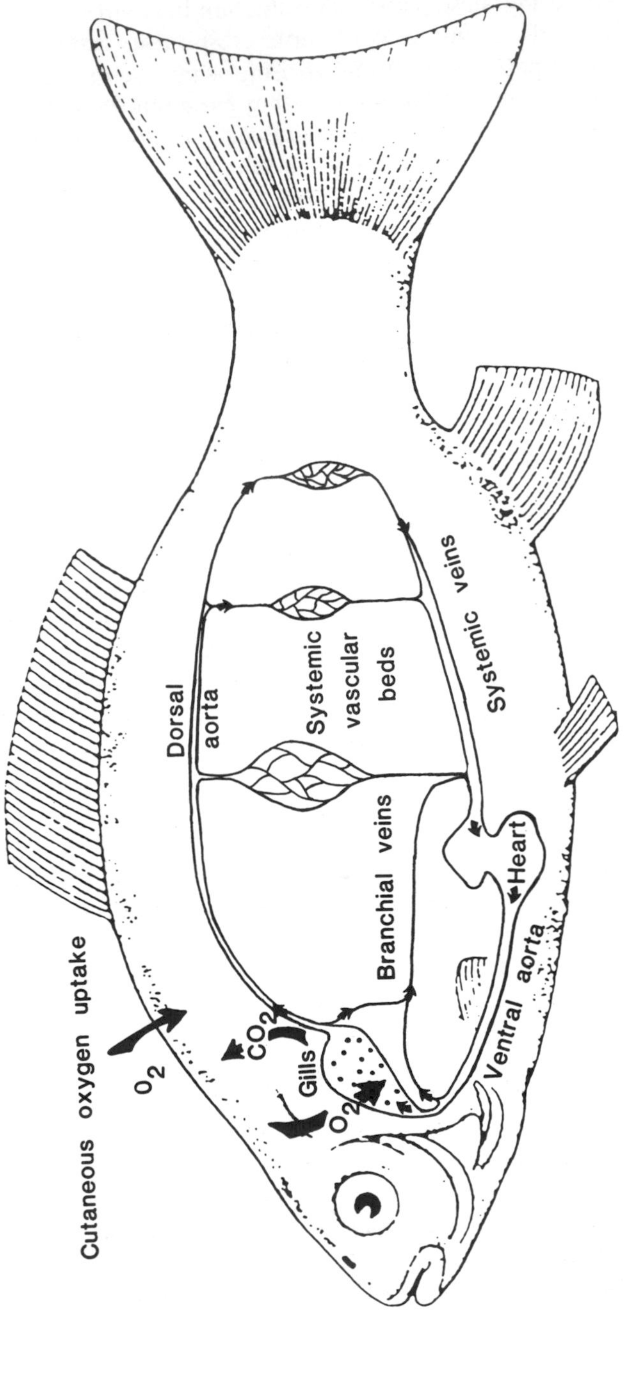

Figure 3.3: The basic scheme of 'single circulation' in fish.

The Heart

Since it is the heart that is responsible for pumping blood around the circulation, any change in either the rate or force of contraction will affect blood flow. Like those of mammals, the hearts of fish possess the ability to contract rhythmically in the absence of any external stimuli. This phenomenon is due to the presence of 'pacemaker' cells within the cardiac tissue which generate a rhythm of action potentials by virtue of their unstable membrane potential. These action potentials propagate rapidly over the myocardium and initiate its contraction. In mammals the dominant pacemaker activity is restricted to a small area of the right atrium known as the sino-atrial (S-A) node. In fish, pacemaker activity seems to be more diffuse and typical nodal tissue is not apparent (Satchell 1971). The rate of pacemaker depolarisation increases with temperature, thus, since fish do not generally control their body temperature, changes in ambient temperature will affect heart rate directly. The intrinsic rhythmic activity of the fish heart is also continuously influenced by neural, humoral (blood-borne) and intrinsic effects which may modify both the rate and force of beating.

The hearts of teleosts, elasmobranchs, and some cyclostomes (but not the aneural heart of the hagfish) receive innervation from the parasympathetic branch of the autonomic nervous system via cardiac branches of the vagus (X cranial) nerve. Activity in the vagus nerve causes the release of the neurotransmitter substance acetylcholine at the heart, which acts via muscarinic acetylcholine receptors (the effect is blocked by the antagonist atropine) and decreases the rate and force of contraction. In resting dogfish (*S. canicula*) at 7°C there is no inhibitory vagal influence acting upon the heart, although at progressively higher temperatures there is an accompanying increase in resting vagal tone (Taylor, Short and Butler 1977).

Excitatory sympathetic innervation of the fish heart was for a long time disputed, it being assumed that cardioacceleration was the result of a reduction of the inhibitory tonus of the vagus nerve. More recently however, sympathetic nerves have been demonstrated in the hearts of some teleosts (Gannon and Burnstock 1969), these adrenergic nerves reaching the heart via the vagosympathetic nerve trunk and the first spinal nerve. In elasmobranchs however, sympathetic innervation is either sparse (Gannon, Campbell and Stachell 1972), or lacking altogether, and there appears to be no adrenergic innervation involved in the control of the heart (Short, Butler and Taylor 1977).

In addition to nervous regulation of the heart, numerous blood-borne substances are known to affect cardiac function in vertebrates, among these are glucagon, thyroxin, 5-HT (serotonin) and the catecholamines, adrenaline and noradrenaline. In fish it has been shown that the levels of circulating catecholamines increase in response to hypoxia (Butler *et al.* 1978; Butler, Taylor and Davison 1979) and physical stress (Nakano and Tomlinson 1967), and it is the effects of these catecholamines upon the circulation of fish that have received the most attention to date. In man and other mammals, the actions of circulating catecholamines on the heart are now accepted to be of little significance in normal health (Rushmer 1970). However, in fish their role is believed to be of much greater importance; this may be particularly true of elasmobranchs in which adrenergic innervation appears to be absent.

Adrenaline and noradrenaline may pass into the bloodstream either by direct secretion from the chromaffin tissue in which they are stored (the adrenal medulla in mammals), or by diffusion from adrenergic nerve endings. These hormones have their effects upon the heart and other tissues by combining with specialised regions ('receptor sites') on the membranes of 'effector cells' which then become activated. The detailed mechanism of receptor site activation is yet to be fully understood, and will not be dealt with here. Pharmacological studies in mammals have revealed a number of distinct types of catecholamine receptor in the circulatory system, namely α, β_1 and β_2. Similar receptor types have also been identified in the circulatory systems of other vertebrates including fish (Stene-Larsen 1981). Generally, adrenaline and noradrenaline cause increases in the force and rate of contraction of the teleost, heart acting via β-(probably β_2-) adrenergic receptors (adrenoceptors) (Falk, Von Mecklenburg, Myhrberg and Persson 1966; Holmgren 1977). In elasmobranchs, adrenaline increases and noradrenaline decreases the rate of beating, while both these hormones increase the force of contraction (Capra and Stachell 1977). In cyclostome fish these hormones appear to have no effect upon the activity of the heart.

The effects of circulating hormones and nerves upon the heart constitute extrinsic control, but the heart is also influenced by intrinsic mechanisms which arise from the properties of the myocardium itself. Cardiac muscle possesses the property that the more it is stretched, the more vigorously it will contract. Thus, within physiological limits, an increase in the return of venous blood to the heart will increase the degree to which the cardiac muscle is stretched during diastolic filling. This increased stretching will result in a more vigorous systolic contraction and so an increased cardiac stroke volume (the volume of blood pumped per heart beat); this is known as the Frank—Starling relationship. Consequently a decrease in heart rate does not always result in a decrease in cardiac output since the increased diastolic filling time increases the filling of the heart (and so the degree of stretching) resulting in an increased stroke volume. This intrinsic effect upon the heart has been observed in all groups of fish.

Another intrinsic response of the heart associated with an increase in venous return is the Bainbridge effect. It is believed that increased atrial filling stretches the pacemaker, resulting in an increase in its activity, this in turn causes an increase in heart rate. This effect is thought to be of particular importance in the control of the aneural heart of the hagfish. These intrinsic responses of the heart automatically match cardiac output to venous return by varying both heart rate and stroke volume without the need to involve the central nervous system.

The Branchial Circulation

For many years it was a popular belief among biologists that in fish all cardiac output flowed to the systemic circulation via the counter-current gas-exchange surface of the secondary lamellae in the gills (see Figure 3.1). This simple model for blood flow was questioned when, in 1964, Steen and Kruysse showed anatomically in the eel that deoxygenated blood flowing from the heart could bypass the secondary lamellae and pass from the afferent filament artery to the efferent filament artery via a central compartment in the gill filament (the central venous sinus). Their observations led them to suggest that the number of secondary lamellae that are perfused with blood can be varied, and that the pattern of blood flow within the gills might be controlled so that, within physiological limits, the functional surface area (i.e. the number of secondary lamellae perfused) of the gills might be regulated to provide just enough oxygen to maintain aerobic metabolism and yet minimise any undesirable flux of salts or water between the blood and the aquatic environment.

In the intervening years the precise model for blood flow proposed by Steen and Kruysse (1964) has been questioned, and doubt has been cast on whether controlling lamellar perfusion will affect the flux of salts and water to any great extent, since such fluxes will occur whether the blood in each lamellae is flowing or stationary (Booth 1979). Even so, the concept of a variable functional surface area has remained (see later). Subsequently numerous anatomical studies of the microvasculature of the gills of many fish species have revealed a very complex structure. Although it is almost certain that a major proportion of cardiac output does indeed pass across the gas exchange surface in the gills and then passes directly on to the systemic circulation, it is now apparent that alternative blood pathways exist within the gills of many fish species and that even if cardiac output remains unchanged, the functional surface area of the gills may be varied by a redistribution of blood flow. It is not possible to describe here the great variety of gill vascular arrangements that exist, since there are often major differences even between closely-related species. Discussion will be restricted to the principle routes via which blood may traverse the gills in addition to the direct route from the heart to the systemic circulation via the gas exchange surface. It should be stressed at this point that not all these possible alternative blood pathways have been found in all species of fish.

In some species (e.g. the eel and the dogfish) it appears that deoxygenated blood flowing in the afferent filament artery towards the secondary lamellae may pass via various possible routes, directly back to the venous circulation instead of entering the secondary lamellae. Such alternative blood pathways could be of particular importance in permitting the control of the number of lamellae perfused without having to change cardiac output. In addition, newly oxygenated blood leaving the secondary lamellae may also enter the venous circulation via anastomoses (small connecting vessels) between the efferent filament artery and the central venous sinus (see Figure 3.4). Although this alternative route will not affect lamellar perfusion directly, it will obviously affect the flow of oxygenated blood to the systemic circulation. In contrast to Steen and Kruysse's original model for blood flow, it now seems that there are no functional blood shunts connecting the afferent and efferent filament arteries, except perhaps via the basal channels in the secondary lamellae (see later). It should be appreciated that the central venous sinus is connected to the branchial veins and is, therefore, at a pressure close to central venous pressure. Consequently blood can only flow *into* the central venous sinus from either the afferent or efferent filament arteries. It is worth noting, however, that any shunting of blood past the secondary lamellae, and/or any cutaneous uptake of oxygen, may invalidate the Fick method as a means of calculating cardiac output (Metcalfe and Butler 1982).

From the foregoing discussion it might be expected that fish would

Figure 3.4: A diagrammatic representation of possible blood flow routes in the gill. Deoxygenated blood flowing from the heart to the gills enters the afferent filament artery (AFA) from which it flows into the secondary lamellae (SL) via the afferent lamellar arterioles (ALA). In some species however some proportion of the blood flowing in the afferent filament artery may bypass the secondary lamellae by returning to the venous circulation, which is illustrated here by the flow of blood via afferent arteriovenous anastomoses ($AVA_{(aff)}$) and central venous sinus (CVS) to the branchial vein (BV). After gaseous exchange and osmoregulation blood leaves the secondary lamellae and enters the efferent filament artery (EFA) via efferent lamellar arterioles (ELA). The major proportion of blood in the efferent filament artery then flows on to the systemic circulation. However in many fish species oxygenated blood leaving the secondary lamellae may pass to the venous circulation which is illustrated here by the flow of blood from the efferent filament artery to the central venous sinus via efferent arteriovenous anastomoses ($AVA_{(eff)}$).

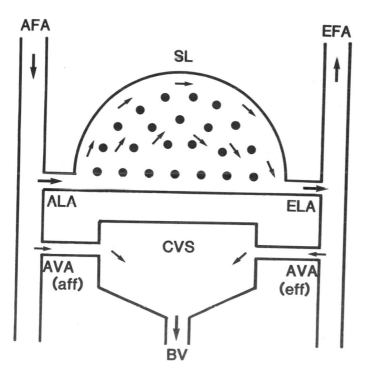

control the distribution of blood flow through their gills. Our knowledge of the control of branchial perfusion comes basically from two lines of experimentation; pharmacological and nerve stimulation studies on isolated, perfused heads, gills and whole fish, and from physiological experiments in which the responses to stimuli such as hypoxia and exercise have been observed. The latter will be dealt with towards the end of this chapter. From the numerous pharmacological studies, it appears that adrenaline and noradrenaline produce an overall decrease in vascular resistance to blood flow through the gills by dilating the blood vessels; this is mediated via β-adrenergic receptors. Alpha-adrenergic receptors also appear to be present in the gill vascular bed but generally the constrictor response to α-receptor stimulation is obscured by the dominant β-receptor dilatation. Acetylcholine has a consistent constrictor effect on the gill blood vessels which is mediated via muscarinic receptors. Among the various models that have been proposed for the actions of vasomotor substances on blood pathways through the gills, there appears to be agreement that the dilator actions of catecholamines preferentially direct blood across the gas exchange surface by recruiting previously unperfused secondary lamellae. In trout, catecholamines increase both the number of perfused secondary lamellae and hence the functional surface area (Booth 1979; Bergman, Olson and Fromm 1974). In eels, catecholamines increase the oxygen tension of post-branchial blood (Peyraud-Waitzenegger 1979) and increase the proportion of cardiac output that reaches the systemic circulation (Hughes, Peyraud, Peyraud-Waitzenegger and Soulier 1981). In addition to these effects, it seems that catecholamines may also increase the permeability of the gill epithelium to oxygen. In trout, adenosine has been shown to constrict the branchial arteries. This purine is known to play a role in the control of blood flow in mammals, but the significance of its effects in fish is not known.

In intact fish, catecholamines may reach the effector cells in the blood vessels via either humoral or nervous routes, while presumably acetylcholine reaches the effector cells only via nerve endings since it does not circulate in the blood. However, the actions of acetylcholine alone cannot be regarded as evidence for cholinergic innervation of the gill blood vessels since vasomotor responses to acetycholine can occur in vascular smooth muscle even if it receives no cholinergic innervation. Little has been reported concerning the nervous control of blood flow in the gills of fish. In Atlantic cod (*Gadus morhua*) both sympathetic (adrenergic dilator and constrictor) and parasympathetic (cholinergic constrictor) vasomotor innervation appear to be present; these nerves reaching the gills via the vagosympathetic nerve trunk (see Chapter 1). Studies conducted on the dogfish (*Scyliorhinus canicula*) in our laboratory seem to indicate that there is no direct vasomotor innervation of the gill blood vessels in this species and presumably all branchial vascular control is exerted via circulating hormones or intrinsic or mechanical mechanisms. The sites of cholinergic vasoconstriction have not yet been fully identified, but the balance of evidence suggests that these are post-lamellar, probably in the efferent filament artery close to its junction with the efferent arch artery, at least in teleosts.

In addition to the nervous and humoral control of the branchial circulation described above, there is some indication that the distribution of blood flow might also be controlled by intrinsic mechanisms in response to changes in oxygen tension. Experiments on some teleost and an elasmobranch species have shown that a reduction in the oxygen tension of the aquatic environment and/or the fluid perfusing the gills, induces constriction of the gill blood vessels. Although the sites of the vasoconstriction are as yet unknown, it is believed that they are post-lamellar, either in the efferent lamellar arteriole or in the efferent filament artery. The functional significance of this vasoconstriction is unclear, it may facilitate a redistribution of blood flow away from areas of the gill where there is localised hypoxia (a similar vasoconstriction in response to localised hypoxia in the lungs of mammals is well known), or, if there is general hypoxia, it may result in an increase in the number of secondary lamellae perfused as a result of an increased afferent perfusion pressure which may dilate previously unperfused lamellae. If this indeed occurs it would serve to increase the functional surface area of the gills and thus the capacity for gas exchange.

Although nervous, hormonal and intrinsic mechanisms may play a part in the control of branchial perfusion, there is a growing body of evidence that suggests that the distribution of blood flow might be regulated simply by changes in blood pressure and pulsatility. Morgan and Tovell (1973) originally suggested that the increase in blood pressures observed in the trout in response to exercise might passively force open previously unperfused secondary lamellae. A more extensive study using isolated gills of the lingcod has demonstrated that manipulating input pressure causes lamellar recruitment. Also, increases in blood pressure have been shown to alter the distribution of blood flow *within* individual lamellae by preferentially directing flow across the central region of the lamellae at the expense of flow through the marginal and basal regions (Randall 1982). This mechanism of intralamellar recruitment would increase the functional surface area of each lamella and hence their capacity for gas transfer.

The gills of fish consist of a large number of repeating gas exchange units, the gill filaments, which are perfused in parallel. This being the case it may be asked whether all the filaments are controlled equally, or whether the perfusion of some filaments, or even whole gill arches, is affected differently by the various control mechanisms described above. This aspect of the control of branchial blood flow has received little attention. However, in one study of the trout, Booth (1979) has shown that both hypoxia and adrenaline have differing effects on various gill arches. In comparison with the normoxic control, in which the proportion of secondary lamellae perfused is about 60 per cent in all four gill arches, both hypoxia and adrenaline decrease the proportion of lamellae perfused on the anterior hemibranch of the first gill arch, while there is a marked increase in the proportion of lamellae perfused in the last gill arch, with an overall increase in lamellar perfusion of 29 per cent in response to adrenaline and 22 per cent in response to hypoxia.

The Systemic Circulation

Our knowledge of the control of the distribution of blood flow between the various vascular beds of the systemic circulation in fish is restricted mainly to pharmacological studies. From these it has been possible to describe the responsiveness of the systemic circulation to neurotransmitters and to hormones. However, little has been reported about the regulation of blood flow *in vivo*, or about any changes that occur in blood flow distribution in response to various stimuli.

Oxygenated blood leaving the gills enters the dorsal aorta, from which it is distributed to the various organs and tissues of the systemic circulation. As was pointed out earlier, not all cardiac output reaches the dorsal aorta from the gills, since a variable proportion may be directed back to the heart. Consequently systemic blood flow is, to some extent regulated by those mechanisms which control blood flow within the gills. However, here we shall deal only with those processes which affect the flow of blood once it has entered the dorsal aorta.

Generally, the dominant effect of adrenaline and noradrenaline on the systemic vascular bed is vasoconstriction in both teleosts and elasmobranchs. This effect is mediated via α-adrenergic receptors. In some instances blockade of this α-constriction has revealed a weak vasodilatation mediated via β-adrenergic receptors. This response to catecholamines is the reverse of their effects on the branchial circulation described earlier where the dominant effect is a β-receptor mediated vasodilatation. This difference is presumably due to different receptor type densities, or different vascular sensitivities. The actions of acetylcholine on the systemic vascular bed seem to be more complex than those of adrenaline and noradrenaline. In teleosts, acetylcholine has been shown to have a weak, direct, vaso-constrictor effect which is mediated via muscarinic receptors. In addition there appears to be an indirect effect which is mediated via nicotinic receptors and causes the release of adrenaline, this then causes constriction via α-adrenergic receptors. In the trout there is a further constrictor response to acetylcholine mediated via nicotinic receptors. This effect may either be direct, via nicotinic receptors in the vascular smooth muscle (however no such vasoconstriction mediated directly via nicotinic receptors is known in mammals), or an indirect effect via the release of some unknown transmitter, possibly ATP (this response is closely mimicked by the actions of ATP) (Wood 1977). Unlike the situation in mammals, there appears to be no dilator response to acetylcholine in the systemic vascular bed of fish.

Nerve stimulation studies on various vascular beds or arteries of the systemic circulation of teleosts have demonstrated that the dominant effect is constriction mediated via α-adrenergic receptors. Weak β-adrenergic dilator and cholinergic constrictor responses have also been reported in some instances. At present there is some dispute as to whether the adrenergic control of the systemic circulation is mediated dominantly via circulating catecholamines or via adrenergic nerves in teleosts (Smith 1978) and it seems that different species may differ in this respect. In elasmobranchs, recent studies suggest that adrenergic control of the systemic circulation is exerted only via circulating catecholamines and that there is no direct nervous control (Opdyke, Holcombe and Wilde 1979).

Vascular Sensory Systems

Assuming that fish do indeed regulate the distribution of blood flow in response to various stimuli it is to be expected that they would possess sense organs which monitor the internal and external environments and can transmit afferent information to the central nervous system.

In mammals, blood pressure is regulated through the carotid and aortic baroreflexes. An increase in blood pressure stimulates baroreceptor activity, this evokes a reflex adjustment of cardiac output and peripheral resistance so that normal blood pressure is restored. This knowledge of mammals has led to the search for a similar baroreceptor reflex in fish. A number of studies in both teleosts and elasmobranchs have demonstrated that an increase in blood pressure brings about a decrease in heart rate by way of increased efferent activity in the cardiac vagal nerves (Jones and Milsom 1982). Irving, Solandt and Solandt (1935) have demonstrated that increasing the blood pressure in the ventral aorta of elasmobranchs causes bursts of afferent nervous activity in the branchial branches of the vagus nerve. However, despite such evidence, it has been suggested that, at least in elasmobranchs, the circulatory system acts as a simple elastic reservoir and that there is no active control of blood pressure (Opdyke *et al.* 1979).

Numerous studies have demonstrated a coordinated response of the respiratory and cardiovascular systems of fish to a reduction in environmental oxygen tension (see later). This suggests that fish are able to monitor external and/or internal oxygen levels. In mammals the existence of peripheral oxygen receptors located in the carotid and aortic bodies is well known. In addition, a decrease in arterial oxygen tension directly activates cardioinhibitory areas in the brain. Homologous chemoreceptors have been identified in birds and amphibians, but the situation in fish is less clear. It appears that there are superficially placed oxygen receptors in fish which detect a decrease in environmental oxygen tension. In trout Daxboeck and Holeton (1978) have recently shown that these are located on the anteriodorsal surface of the first hemibranch and are innervated by the IX and/or X cranial nerves. These receptors appear to mediate the decrease in heart rate, but not the increase in ventilation (see earlier) observed in response to hypoxia in this species. In dogfish, similar superficial oxygen receptors are reported to be spread diffusely over the mouth and snout (Butler, Taylor and Short 1977).

Coordinated Activity of the Respiratory and Circulatory Systems

Cardio-respiratory Coupling

Many workers have suggested that a reasonably fixed relationship exists between heart beat and respiratory movements in fishes (see Hughes 1972). Certainly for short periods the two rhythms may be closely coupled, but in the dogfish, *S. canicula*, this may occupy less than 5 per cent of 24 hours of recording. Complete synchrony occurs in the trout during hypoxia, when heart rate declines, although this is not so for all fish. Nonetheless, there are bursts of efferent activity in the cardiac branch of the vagus of the dogfish, *S. canicula*, which are in phase with respiratory movements (Figure 3.5a) and which appear to arise reflexly from proprioreceptors in the pharynx. These bursts may loosely couple the respiratory and cardiac pumps, thus matching the pulsatile flows of water and blood for optimum gas exchange (Taylor and Butler 1982).

Hypoxia

Although environmental hypoxia is not normally encountered by pelagic marine fish or those living in fast flowing streams and rivers, experimentally induced hypoxia does give an insight into aspects of respiratory and cardiovascular control. Oxygen uptake and arterial PO_2 can be maintained, during hypoxia, by increasing ventilation, reducing the partial pressure difference of oxygen across the gills and decreasing PO_2 in venous blood (Jones, Randall and Jarman 1970). In rainbow trout, ventilation increases progressively as PO_2 falls, reaching 13 times the normoxic value at a PO_2 of 5.3 kPa. Both frequency and amplitude of breathing increase and this extra activity means that the respiratory muscles themselves use proportionately more of the diminishing oxygen supply. The relative amount of oxygen extracted from the water (percentage extraction) decreases in trout during hypoxia, but in carp, where ventilation increases to approximately 8 times the normoxic value, percentage extraction of oxygen from the water does not change (Lomholt and Johansen 1979). Carp are, in fact, noted for their extreme tolerance of hypoxia (see earlier). Arterial PO_2 can only be maintained by increasing ventilation until a certain environmental PO_2 is reached (this depends largely on the oxygen affinity of the blood). As arterial PO_2 falls, oxygen uptake can then be maintained by reducing venous PO_2. The trout is able to maintain its oxygen uptake at 15°C as environmental

Figure 3.5: Recordings of efferent activity from the left branchial cardiac branch of the vagus nerve in the dogfish (*Scyliorhinus canicula*). (a) Activity related to respiratory movements. Contractions of the walls of the pharynx (upward deflection) are associated with a burst of action potentials in the cardiac vagus. Smaller action potentials are present between and within the bursts. (b) Activity in response to ventilation of the fish with normoxic and hypoxic water. The bursts of activity are slower and contain fewer action potentials during hypoxia, whereas the activity in the continuously firing units is markedly increased.

(a)   (b)

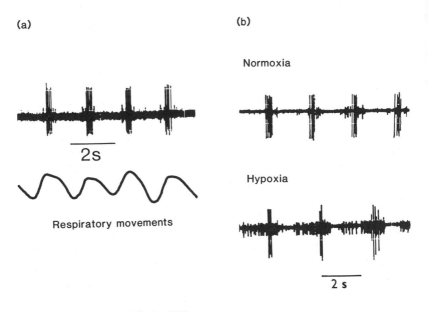

Normoxia

2s

Hypoxia

Respiratory movements

2 s

Modified from Taylor and Butler 1982

PO_2 falls to 5.3 kPa, whereas the dogfish (*S. canicula*) is not (Butler and Taylor, 1975). This is, no doubt, because the dogfish is unable to increase its ventilation volume by anything like the same proportion as the trout. Work in our laboratory indicates that the dogfish is particularly sensitive to operative procedures and may have to be left several days before it reaches a resting condition. Previous studies on the effects of hypoxia in dogfish may have been performed on already stressed and hyperventilating animals. In trout and tench, TO_2 increases during hypoxia by approximately 4 times, indicating an increase in effective surface area of the gills (lamellar recruitment), or a decrease in diffusion distance, or both. Such a change is not apparent in the dogfish.

A feature common to both elasmobranchs and teleosts during environmental hypoxia is a reduction in heart rate (bradycardia) mediated by an increase in vagal activity (Figure 3.5b). The environmental PO_2 at which this bradycardia occurs varies between species, but within the same species it is also affected by temperature (as is the reduction in oxygen uptake) (Butler and Taylor 1975). The physiological significance of the bradycardia is not clear, particularly as it is accompanied by an increase in cardiac stroke volume which maintains cardiac output constant (thus, as ventilation volume increases during hypoxia, the ratio, $\dot{V}_w/\dot{V}_b$ increases. It has been calculated that the power output of the heart decreases slightly during the bradycardia of hypoxia, and it has been demonstrated in the dogfish (*S. canicula*) that a profound bradycardia may function to maintain arterial oxygen tension at as high a level as possible (Taylor *et al.* 1977). The increased ventral aortic pulse pressure associated with the bradycardia may be important in producing inter- and intra-lamellar recruitment and in reducing diffusion distance, thus maintaining gas transfer (see earlier). These mechanical effects may be particularly important in the trout where mean blood pressures also increase.

The role of circulating catecholamines, which are known to increase during hypoxia in dogfish (Butler *et al.* 1978), is not that clear. The increase in cardiac stroke volume during hypoxia is the result of the Frank—Starling relationship in dogfish and not of increased concentration of catecholamines (Short *et al.* 1977). Injection of adrenaline causes an increase in arterial PO_2 in the normoxic eel (Peyraud-Waitzenegger 1979), but assuming that blood transit time through the secondary lamellae is slow enough to allow oxygenation (Randall 1982), the only way in which this increase in PO_2 can be achieved under normoxic conditions is by the reduction of an effective afferent to efferent shunt. Movement of blood away from the basal channels, which may be buried in the surface of the filament (Figure 3.1c), into the central channels of the secondary lamellae (intra-lamellar recruitment) could cause such a change. Whether this could be the direct effect of adrenaline, or the indirect effect of changes in cardiovascular dynamics having mechanical effects on the gill lamellae remains to be seen, as important cardiovascular variables were not measured and reported in the experiments on the eels. Similar increases in arterial PO_2 in response to adrenaline injections have been reported in isolated, saline perfused heads of cod (Pettersson and Johansen 1982). These results have to be treated with caution, because it is known that saline causes gross morphological changes (which increase diffusion distance and reduce TO_2) in the gills of some fishes, whereas these changes are prevented or partially reversed by adrenaline (Pärt, Tuurala and Soivio 1982).

Exercise

During aerobic exercise there is an increased demand for oxygen which can only be fully met by an increase in blood flow to the active muscles, where massive vasodilatation occurs (presumably in response to metabolic products) and where more than 90 per cent of the extra oxygen is consumed. Blood flow to the rest of the body, e.g. the gut, is reduced, probably as a result of sympathetic nervous activity causing vasoconstriction via α-receptors (Randall 1982). However, the major cardiovascular change is an increase in cardiac output, which results mainly from an increase in stroke volume (Jones and Randall 1978). In fact, for the trout, a 7.7 times increase in oxygen uptake at maximum sustainable swimming speed is met largely by a 2.5 times increase in the difference between arterial and venous O_2 content and a 2.2 times increase in cardiac stroke volume. The increase in heart rate makes only a minor contribution. A similar situation also exists in the swimming dogfish, although oxygen uptake was only 1.75 times resting, in this case (Piiper, Meyer, Worth and Willmer 1977). Whether the veins of fishes are sufficiently well-valved or compressed adequately during swimming to augment venous return is not certain, and the influence of the Frank–Starling mechanism upon cardiac stroke volume during exercise is thought to be less important than neurohumoral factors (Jones and Randall 1978). Mean blood pressure and pulse pressure in the ventral aorta increase during exercise in the trout, whereas changes in the dorsal aorta are much smaller. With the increase in cardiac output that occurs, there is clearly overall vasodilatation in the systemic circuit. Mean pressures in dorsal and ventral aortae do not change during moderate swimming activity in the dogfish.

In the trout, ventilation during maximum sustainable exercise increases to 8 times resting, which is greater than the rise in cardiac output so that, during hypoxia, $\dot{V}_w/\dot{V}_b$ increases. Despite the decreased transit time of water through the gills, the amount of oxygen extracted from the water does not decrease. Although cardiac output increases by 3 times, residence time of the blood in secondary lamellae may only fall to half the resting value, if lamellar recruitment occurs (Jones and Randall 1978). Arterial PO_2 is maintained close to the resting value in the trout during the full range of aerobic exercise and overall TO_2 increases by 6 times (although it may not be as great as this at the lamellae — see Jones and Randall, 1978). Similar mechanisms to those discussed earlier (i.e. during hypoxia) are, no doubt, operative to cause the rise in TO_2 during exercise. Although it is assumed that the levels of circulating catecholamines rise during exercise (Randall 1982) this has not been clearly demonstrated, as the only reports on the subject involved grasping the tail of the fish (Nakano and Tomlinson 1967; Opdyke, Carroll and Keller 1982).

An interesting aspect of ventilation during exercise in some fish, e.g. skipjack tuna (*Katsuwonus pelamis*), mackerel (*Scomber scombrus*) and bluefish (*Pomatomus saltatrix*), is the use of ram ventilation; simply moving forward with the mouth open. When forced to swim at increasing velocities, it has been found that these fish switch from active to ram ventilation within a certain velocity range. Ventilation of the gills is not then merely controlled by swimming velocity, but also by the gape of the mouth. The nature and location of the receptors involved in the switch to ram ventilation and in the control of mouth gape are not known (Jones and Randall 1978; Randall 1982), although those on the gills which inhibit ventilation (see earlier) could be important. The functional significance of ram ventilation is that it improves overall efficiency by producing streamlined flow over most of the length of the fish, thereby improving its hydrodynamic performance (Freadman 1981). So, in very active fish the function of ventilating the gills is taken over largely by the locomotor muscles and, because of the improved efficiency of locomotion, the demands on the circulatory system are reduced.

Acknowledgement

The authors' work on this topic is supported by the Science and Engineering Research Council.

4 OSMOREGULATION AND THE CONTROL OF KIDNEY FUNCTION

J.C. Rankin, I.W. Henderson and J.A. Brown

Fish kidneys make a negligible contribution to the excretion of nitrogenous waste, which mainly passes out across the gills as ammonia or ammonium ions (Goldstein 1982), in close association with the mechanisms which regulate monovalent ion uptake and acid-base balance (Evans 1982). Renal excretion is primarily concerned with either water (in freshwater fish) or, in marine fish, divalent ion (mainly magnesium and sulphate) elimination. The kidney shares the responsibility for regulating body fluid pH with the gills (Kobayashi and Wood 1980; McDonald and Wood 1981) but since it probably plays a minor role (Cameron 1980) this aspect will not be discussed further. To put the principal functions in perspective the main features of fish osmoregulation will be briefly summarised. (For further details and references see Rankin and Davenport 1981.)

Fish Osmoregulation

In this chapter four broad categories of the vast array of enormous phylogenetic diversity which can loosely be classed as 'fish' will be considered; the agnathous cyclostomes, the bony actinopterygians (in particular the teleosts), the cartilaginous elasmobranchs and the air-breathing dipnoans. It should be emphasised at the outset that the genealogies of these groups are complex and extant representatives are extremely remote cousins.

The watery environments of fish range from fresh waters which may have almost no dissolved solutes, through brackish waters to the sea, which on average has an osmolarity of about 1 osmol^{-1}, and hypersaline brine lakes. In the face of the great differences in solute and water availability in their environments, fish generally maintain a constancy of their internal body fluid composition via a process termed osmoregulation. To achieve this 'disequilibrium' between the ambient water and their extracellular fluids they employ a number of homeostatic organs that include gill, kidney, urinary bladder, gut and, in elasmobranchs, the rectal gland.

The most primitive 'fish', the myxinoids, have extracellular fluids resembling sea water in ionic composition, although serum magnesium is maintained at about one tenth of the sea water concentration and serum sodium is correspondingly increased (Bellamy and Chester Jones 1961), since the body fluids are iso-osmotic with sea water.

All other fish are descended from freshwater ancestors which, in the course of evolution, acquired reduced body fluid salt concentrations which alleviated the osmotic (water entry) and ionic (salt loss) problems associated with life in a hypo-osmotic environment. However, once their bodies had become adapted to these lower salt concentrations they appear to have been obliged to retain them if they subsequently re-colonised the marine environment. This led to problems of osmotic loss of water and diffusional entry of salts, problems which were solved in two ways by different groups of fish.

Elasmobranchs and, as far as we can tell from the sole surviving species, crossopterygians, evolved high extracellular urea and trimethylamine oxide concentrations to bring their body fluid osmolarities up to the sea water level. (The plasma is in fact very slightly hyperosmotic to the environment, permitting a small passive water influx to balance renal and other losses.) The integument is relatively impermeable to salts to allow limitation of diffusional entry, and a sodium chloride excretory organ, the rectal gland, has evolved to assist in the elimination of excess dietary sodium. The high body fluid osmolarity would have exacerbated the osmotic problems of those species which attempted to re-invade freshwater and consequently very few freshwater representatives of groups of fish which have adopted this osmoregulatory strategy are found.

Those bony fish and lampreys which re-evolved the ability to live in sea water maintained the reduced body fluid osmolarities characteristic of freshwater fish. This meant that they were faced with the problem of osmotic loss of water, as well as diffusional entry of salts. The only way this loss could be replaced was by drinking sea water, which could not however be absorbed directly because of the adverse osmotic gradient. The gradient must first be reduced by diffusion of sodium chloride into the body fluids across the wall of the oesophagus and some further equilibration in the stomach to the point where solute-linked water uptake (i.e. dependent on active salt uptake) could occur in the intestine. A considerable additional salt burden is thus incurred as an unavoidable consequence of the replacement of osmotic water loss. Excess sodium and chloride ions are excreted by chloride cells in the gills, but magnesium and sulphate ions, which diffuse down their electrochemical gradients into the fish, are removed renally. The osmotic and ionic problems faced by a teleost fish in freshwater and in the sea, together with the measures taken to counteract them, are shown in Figure 4.1.

Most species of fish are stenohaline, that is they are unable (in contrast to euryhaline species) to withstand large changes in environmental salinity. A few species, however, embark on a series of quite remarkable journeys between freshwater and the sea during their life histories. The catadromous (to spawn in the sea) and the anadromous (to spawn in freshwater) migrations of salmon and lampreys are familiar examples. Other species inhabit brackish water and here daily or seasonal salinity changes may impose equivalent but more acute problems.

The activities of the osmoregulatory systems outlined in Figure 4.1 are tempered by the neuroendocrine system. Knowledge of direct neural effects on renal and extrarenal function is poorly described at present and, with one or two notable exceptions, information on hormone actions is largely derived from comparative pharmacological, rather than physiological, studies. In this context fish possess the typical vertebrate endocrine make-up, but in some groups unique features, such as the corpuscles of Stannius and the urophysis, exist.

In the face of considerable variations in the aquatic environment, fishes, especially teleosts, exhibit an enormous adaptive radiation into virtually all lakes, rivers and oceans by virtue of their ability to regulate their body fluid volumes and compositions to remedy excesses and deficiencies of solutes and water in their habitats.

Kidney Function in Cyclostomes

Hagfish

Hagfish have primitive kidneys, with a small number of large, segmentally arranged glomeruli connected by extremely short tubules to the urinary ducts (Fänge 1963). Little is known about the function of the kidney, let alone about any possible regulatory mechanisms. It may be involved in magnesium ion excretion (Morris 1965), but the evidence for this is conflicting (Alt, Stolte, Eisenbach and Walvig 1981). It has been speculated that the Atlantic hagfish

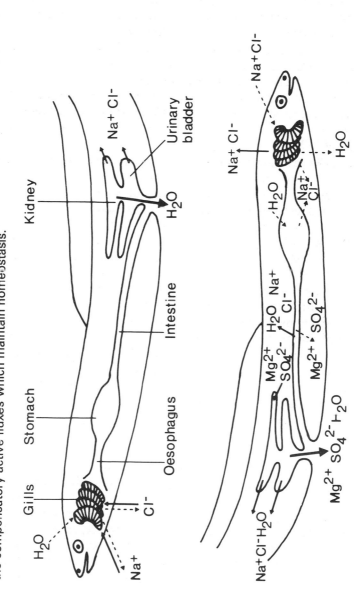

Figure 4.1: Diagram to illustrate the fluxes of water and salts in a freshwater (above) and marine (below) teleost. Broken arrows represent passive fluxes between the body fluids and the environment. Solid arrows represent the compensatory active fluxes which maintain homeostasis.

Myxine may be able to increase urine flow rate in response to osmotic water influx caused by slight reductions in environmental salinity (Fänge 1963), and the fact that glomerular filtration rate (GFR) is very sensitive to arterial pressure changes in both the Pacific hagfish *Eptatretus* (Riegel 1978) and *Myxine* (Alt *et al.* 1981) suggests the possible existence of a simple volume regulatory mechanism whereby water entry leads to increased blood volume and pressure, which increases filtration and urine flow rate. The vertebrate kidney may thus have originated as a volume regulatory organ in the marine protovertebrates, acquired its osmoregulatory functions with the colonisation of freshwater and its nitrogenous excretory function with the emergence from the aquatic into the terrestrial environment.

Lampreys

Lampreys have probably been evolving separately from hagfish for hundreds of millions of years and their osmoregulatory mechanisms show striking parallels with those of teleost fish (Morris 1972; Youson 1982). Recent work on the euryhaline river lamprey, *Lampetra fluviatilis*, has taken advantage of the fact that its kidney structure renders it especially convenient for micropuncture studies (Logan, Moriarty and Rankin 1980), so the renal physiology of this species will be considered in some detail.

River lampreys are able to maintain almost constant plasma composition when adapted to either freshwater or sea water (Figure 4.2) although once they have entered freshwater on their spawning migration they rapidly lose the ability to osmoregulate in sea water, possibly as a result of degeneration of the intestine under the influence of gonadal steroids (Pickering 1976). In common with all vertebrates having freshwater ancestors the number of nephrons has shown a large increase over that found in the presumably primitive segmental arrangment of hagfish. Some idea of the number of functional nephrons can be obtained by dividing the whole animal glomerular filtration rate (GFR) by the mean single nephron filtration rate (SNGFR). This calculation is subject to some uncertainty due to errors caused by some reabsorption of the ^{3}H-inulin used as an indicator of filtration rates, and is based on the somewhat optimistic assumption that the few nephrons from which it is possible to sample in each animal are representative of the entire population, but 36 g lampreys would appear to have 2300-2800 functional glomeruli (Moriarty, Logan and Rankin 1978). This number is required to maintain the GFR of 600 ml kg^{-1} 24 h^{-1} found

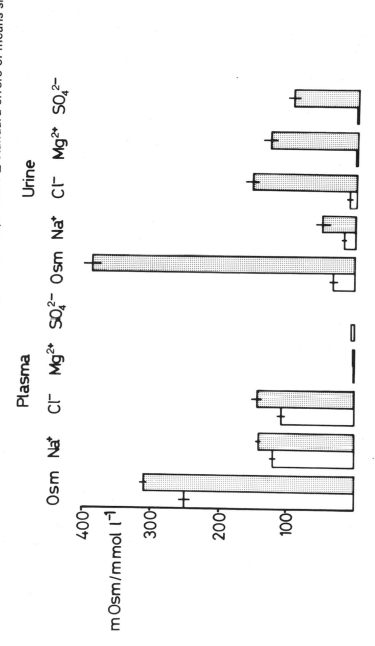

Figure 4.2: Plasma and urine osmolarities and ionic concentrations in the river lamprey (*Lampetra fluviatilis*) adapted to freshwater (open columns) or sea water (stippled columns). Means ± standard errors of means shown.

in freshwater lampreys (Logan *et al.* 1980). Although almost half the filtrate is reabsorbed (probably as an unavoidable consequence of the large osmotic gradient which exists across the walls of the distal tubules and collecting ducts) the mean free water clearance is 29.5 per cent of the body weight per day. Urine volume must be precisely regulated to balance osmotic entry of water across the gills, since the animals are able to maintain an almost constant body weight.

Freshwater lampreys produce a very dilute urine, most of the filtered sodium and chloride ions being reabsorbed in the distal tubule (Logan *et al.* 1980). Sea water adapted lampreys are able, unlike any other fish which have been studied, to produce a urine which is slightly but significantly hyperosmotic to their plasma, with very high concentrations of magnesium and sulphate ions (Figure 4.2). Urine flow rate is only 6 per cent of that of freshwater fish, due to a 70 per cent decrease in GFR and an increase in tubular water reabsorption to almost 90 per cent (Logan, Morris and Rankin 1980). The mechanisms regulating urinary water and salt excretion will be considered separately.

Regulation of Water Excretion.   In hypo-osmotic environments urine flow rate varies with the osmotic gradient across the gills as the kidneys operate to maintain a constant body fluid volume (Figure 4.3). The changes in urine flow produced by variations in external osmolarity up to iso-osmicity are due to changes in GFR, the percentage of the filtered fluid reabsorbed across the tubule wall being relatively constant. The changes in GFR are caused by changes in the SNGFRs of all the nephrons (Rankin, Logan and Moriarty 1980).

Ultrafiltration in the glomerulus occurs because the hydrostatic pressure in the glomerular capillaries exceeds the sum of the hydrostatic pressure in the Bowman's space and the osmotic pressure created by the plasma proteins, which are largely retained by the filtration barrier. Values of these pressures in lampreys before and after transfer from freshwater to approximately iso-osmotic brackish water are given in Figure 4.4. The colloid osmotic pressure rises along the glomerular capillaries as ultrafiltrate is extruded, concentrating the residual protein. In freshwater adapted fish there is clearly always a positive-effective filtration pressure, but after transfer to brackish water filtration equilibrium may well be achieved before the end of the capillary loops. There is no significant difference between renal arterial and glomerular capillary pressures in either freshwater or

Figure 4.3: Relationship between urine flow rate and osmotic gradient between plasma and water (Osm_p — Osm_w) in the river lamprey (*Lampetra fluviatilis*). Means ± standard errors of means shown.

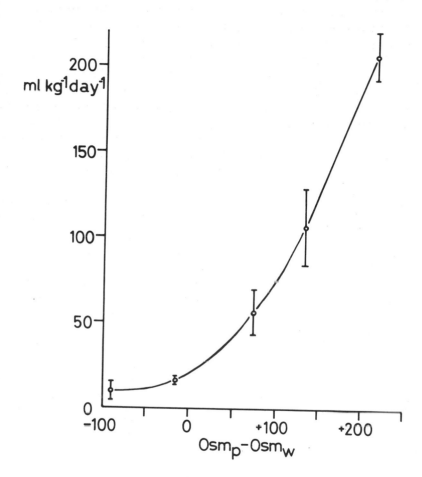

brackish water, but there is a highly significant ($p < 0.001$) difference between glomerular capillary and peritubular capillary pressures in fish in both media, showing that the main sites of vascular resistance in the lamprey kidney are in the efferent arterioles. Although the pressure drop along these vessels is reduced in fish from brackish water, their resistance can be calculated to have actually increased,

Figure 4.4: Diagrammatic representation of a glomerulus and associated structures to illustrate hydrostatic and colloid osmotic pressures in kPa in the renal circulation and nephron of the river lamprey (*Lampetra fluviatilis*) adapted to freshwater (values in parentheses are for animals adapted to approximately iso-osmotic brackish water). Small arrows represent blood flow. Large open arrow represents hydrostatic pressure gradient; large closed arrows represent colloid osmotic pressure gradients.

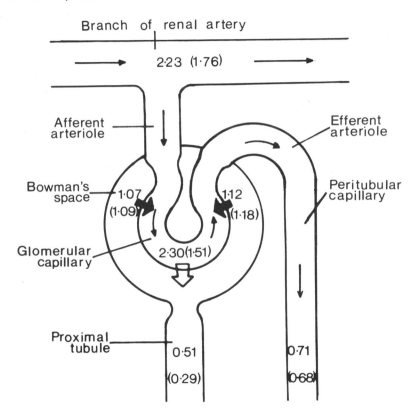

taking into account the 70 per cent decrease in renal blood flow which occurs after transfer (McVicar 1982).

Control of blood pressure and flow seem to be the main determinants of urine flow, but virtually nothing is known about regulation of the circulation in cyclostomes. In teleosts circulating catecholamine levels play a major role (Wahlqvist and Nilsson 1977) and adrenaline is diuretic in the river lamprey (Dalgleish and Rankin, unpubl. obs.). Angiotensin II has not been demonstrated in cyclostomes and so far it has not been possible to demonstrate any effects in lampreys (Rankin and Griffiths, unpubl. obs.). Neurohypophysial hormones have antidiuretic actions in low doses and diuretic actions in high doses in teleosts (Henderson and Wales 1974; Babiker and Rankin 1978) but only diuretic actions of arginine vasotocin (AVT — the only neurohypophysial hormone identified in cyclostomes) have been observed in lampreys (Rankin, Griffiths, McVicar and Gilham 1982).

Our ignorance is equally profound on the stimuli which initiate whatever control mechanisms are involved in the regulation of water excretion. Infusions of hypo-, iso- and hyperosmotic sodium chloride solutions into the body cavity all provoke similar diureses (Rankin and Griffiths, unpubl. obs.) so volume, rather than osmotic, receptors would appear to be involved. Regulation is rapid and precise. Figure 4.5 shows a typical example of the effects of infusing 400 mOsmol l^{-1} sodium chloride solution into the body cavity of a freshwater adapted lamprey (plasma 250 mOsmol l^{-1}) at a rate of 4.5 ml h^{-1} for 1 hour. There was a tenfold increase in urine flow for the duration of the infusion, the extra urine produced in excess of the control rate during the experimental period being 4.37 ml.

Regulation of Salt Excretion.   Freshwater lampreys reabsorb over 90 per cent of the filtered sodium and chloride ions in the distal tubule, to produce a very dilute urine (Logan, Moriarty and Rankin 1980). Salt loading the animal leads to increased sodium chloride excretion. For example infusion of 400 mOsmol l^{-1} sodium chloride solution caused the urine osmolarity of a freshwater lamprey to rise to 300 mOsmol l^{-1}, accounted for mainly by sodium and chloride ions (Figure 4.5). This was not the result of saturation of the tubular salt reabsorption mechanisms, since infusion of distilled water to produce similar or greater diuretic responses caused no rise in urinary osmolarity. However, even hypo-osmotic sodium chloride infusions cause some increase in salt excretion so some unknown mechanism would appear to be regulating distal tubular salt reabsorption in response to the state of salt balance of the animal. Urine from sea water adapted lampreys has lower concentrations of sodium and chloride ions than the plasma (Figure 4.2); the kidney is therefore unable to assist in the elimination of a salt load since any urine production will tend to increase the plasma concentrations of these

Figure 4.5: Urine flow rate (upper graph), urine osmolarity, sodium concentration and (superimposed on sodium concentration) chloride concentration in a freshwater river lamprey, (*Lampetra fluviatilis*) before, during and after an infusion of 400 mOsmol l^{-1} sodium chloride solution into the body cavity.

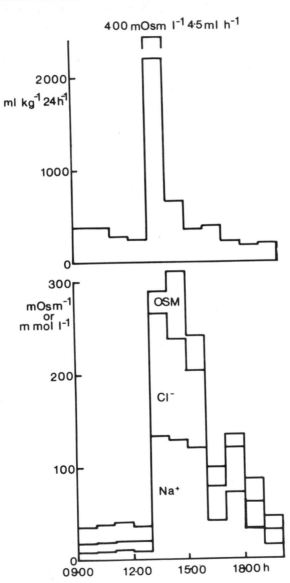

ions, as more water than salt will be excreted (and the lost fluid cannot be replaced without further increasing the salt load).

Excess monovalent ions are in fact excreted by chloride cells in the gills, and the kidney is mainly concerned with secretion of magnesium and sulphate ions (Figure 4.2) which appears to occur in the distal tubules (Rankin *et al.* 1980). No studies have been carried out on the mechanisms of secretion of these ions and their control in lamprey kidneys.

It is possible that increased magnesium ion concentration in the blood perfusing the kidney stimulates magnesium secretion by the tubules, as in the teleost *Lophius* (Babiker and Rankin 1979a), since infusion of magnesium sulphate into a freshwater adapted lamprey causes a large and rapid increase in urinary magnesium concentration (Figure 4.6). This infusion produced a large diuretic response and the amount of magnesium excreted per unit time increased by 2-3 orders of magnitude over control values. In other experiments with lower infusion rates it was possible to demonstrate large increases in urinary magnesium concentrations without increases in urine flow (Rankin and Griffiths, unpubl. obs.). Urinary magnesium concentrations are normally very low in fish in freshwater, but some mechanism must be present to excrete excess dietary magnesium.

Kidney Function in Teleosts

In freshwater, teleosts face the same problem of osmotic water influx across the gills as cyclostomes, and solve it in the same way — by a glomerular kidney in which most of the filtered solutes, but relatively little of the filtered water, are reabsorbed. In sea water, where water is at a premium since osmotic loss across the gills must be replaced by drinking, which leads to a large influx of salts as well as water from the gut, minimal quantities of urine, consistent with the limitations of the urinary concentrating mechanism, are produced. The prime role of the kidney is excretion of divalent ions, particularly magnesium and sulphate, as a result of renal tubular secretory mechanisms. In effect marine teleosts rely on glomerular activity to a minimal degree.

Glomeruli are in general smaller in marine than in freshwater teleosts (Nash 1931); indeed some species (e.g. the angler fish, *Lophius piscatorius*) have aglomerular kidneys consisting only of blind-ending tubules supplied with renal portal venous blood. In such species urine is prepared entirely by secretory processes, which are

Figure 4.6: Urine flow rate (upper graph), urine osmolarity and magnesium concentration in a freshwater river lamprey (*Lampetra fluviatilis*) before, during and after an infusion of 200 mmol l^{-1} magnesium sulphate solution into the body cavity.

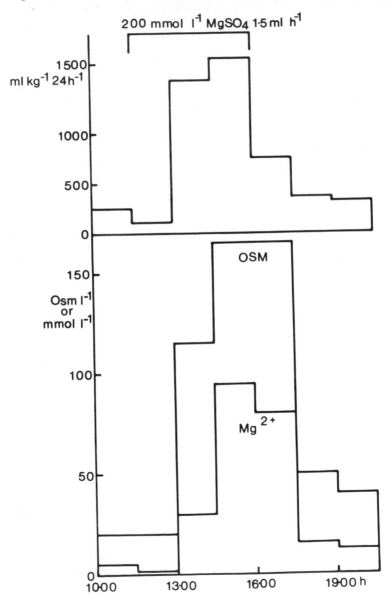

thought to begin by iso-osmotic sodium chloride secretion, with subsequent secretion of magnesium and other divalent ions alongside reabsorption of sodium and chloride. Ion secretion draws water by osmosis into the tubule, urine flow being proportional to the rate of magnesium secretion, which in turn seems to depend directly on the concentration of magnesium ions in the fluid perfusing the kidney vasculature (Babiker and Rankin 1979a). The process thus contrasts significantly with the classical filtration—reabsorption—secretion model so typical of the vertebrate nephron. Even in teleosts with glomeruli the tubules are capable of producing fluid by secretion (Beyenbach 1982).

The lack of glomeruli does not necessarily restrict aglomerular species to the sea, and the goosefish *Opsanus tau* is well able to produce a urine considerably more dilute than its plasma, the urinary bladder playing a significant role in this process (Lahlou, Henderson and Sawyer 1969).

Renal adaptations to varying environmental salinities and their regulation in teleosts have long been a source of fascination for comparative physiologists. In some euryhaline species varying urine production rates may be achieved by regulation of tubular water reabsorption (Oide and Utida 1968; Schmidt-Nielsen and Renfro 1975). In most species however, varying urine production rates parallel changes in overall glomerular filtration rates (Hickman and Trump 1969; Table 4.1). The overall glomerular filtration rate (GFR) is, in this sense, the sum of the filtration rates of all the filtering nephrons. Modifications in GFRs may therefore result from change in individual rates of filtration (single nephron glomerular filtration

Table 4.1: Effects of environmental salinity on urine flow and glomerular filtration in the rainbow trout, *Salmo gairdneri*. Values are means ± standard errors of means, n = number of fish (in brackets), GFR = Glomerular filtration rate, SNGFR = Single nephron filtration rate.

	Freshwater adapted	Seawater adapted
Urine flow $\mu l\ min^{-1}\ kg^{-1}$	76.4 ± 10.4 (21)	5.3 ± 1.4 (5)
GFR $\mu l\ min^{-1}\ kg^{-1}$	140.4 ± 17.2 (15)	18.5 ± 5.8 (5)
SNGFR $nl\ min^{-1}$	1.3 ± 0.2 (5)	3.7 ± 1.1 (3)

rates; SNGFRs) and/or modulation of the size of the filtering population; the latter is termed glomerular intermittency. It is also possible that there are populations of nephrons filtering at different rates, as in mammals, and these can be 'switched on' and 'switched off' according to environmental factors. These various possibilities can only be assessed by studying individual nephron function.

Routine measurement of SNGFR by micropuncture, as carried out in the lamprey, has not yet proved possible in any teleost. An indirect technique has been used, but as yet only in the rainbow trout, *Salmo gairdneri.* ^{14}C-ferrocyanide infused into experimental animals is handled by the kidney in a similar fashion to inulin and its renal clearance thus reflects glomerular activity. An intra-aortic injection of 20 per cent unlabelled ferrocyanide is then administered. This is filtered and a few seconds later the kidney is snap-frozen, removed from the body cavity and the ferrocyanide is precipitated as Prussian Blue by freeze-substitution. Microdissection allows location of the Prussian Blue precipitate in nephrons filtering at the time of the experiment and the radioactivity within the tubule between the glomerulus and the front of the visible precipitate is a measure of the filtration rate (Brown, Jackson, Oliver and Henderson 1978). This technique, whilst not as direct an approach as micropuncture, can be used to measure filtration rates of nephrons deep within the kidney, inaccessible to micropuncture.

Measurement of SNGFRs in this way in the trout indicates that changes in GFR seen during adaptation to different environmental salinities do not coincide with changes in mean SNGFRs. In sea water adapted fish, where total GFR is one tenth of the rate in freshwater adapted fish, mean SNGFR is actually greater than in freshwater adapted fish (Table 4.1).

In several teleost species changes in urine production and GFR mirror changes in the glucose reabsorptive capacity, i.e. the renal tubular transport maximum for glucose (T_{max}) (Lahlou 1966; Hickman 1968). This suggests that filtration and urine production rates are regulated by varying the number of filtering glomeruli, thus varying the number of tubules participating in glucose reabsorption.

Experiments in the rainbow trout have revealed three physiological states in the nephron population; filtering, non-filtering but arterially perfused and non-perfused (Henderson and Brown 1980). The relative proportion of these types varies with environmental salinity (Figure 4.7). In freshwater adapted fish whilst most glomeruli are perfused only 50 per cent filter. In sea water adpated fish 50 per cent of glomeruli are arterially perfused but of these only 10 per cent filter. The haemodynamic factors determining whether a perfused glomerulus does or does not filter are as yet not understood. Overall renal arterial blood flow rates are reduced following adaptation to sea water (Brown and Oliver 1982) but single nephron glomerular blood flow rates and the relative rates in filtering and non-filtering nephrons are not known.

Glomerular filtration rate depends not only on blood flow and effective filtration pressure but also on the ultrafiltration characteristics of the filtration barrier, which is made up of capillary endothelium, basement membrane and an epithelium continuous with Bowman's capsule. In freshwater adapted trout, as in all vertebrates, epithelial podocytes extend as primary processes around the capillary giving rise to interdigitating foot processes, between which lie slit diaphragms. Filtration of small molecules with retention of plasma proteins is believed to be achieved by sieving of large molecules by the basement membrane, the slit diaphragm acting as a final filter preventing penetration of plasma proteins. Water is currently thought to pass across the glomerular capillaries via the endothelial fenestrae, the basement membrane, slit diaphragms and slit pores, i.e. by an essentially extracellular route. Large numbers of slit diaphragms are essential for high water and electrolyte permeability. Preliminary studies suggest that sea water adaptation modifies glomerular ultrastructure of the trout (Brown and Oliver 1982). Epithelial cells and their primary processes and foot processes are flattened and broadened with fewer slit diaphragms, probably resulting in reduced membrane permeability. Thus both structural modifications and reduced blood flows are likely to be associated with glomerular intermittency in rainbow trout.

Control Mechanisms

Control of teleostean kidney function is likely to involve both neural and endocrine factors. Our knowledge of the nervous control of fish kidneys is almost non-existent. Nerve fibres have been observed in association with the renal vasculature and they could well be involved in regulation of glomerular blood flow and filtration pressure and thereby glomerular intermittency and/or SNGFR, but studies on physiological control mechanisms have so far been confined to possible hormonal involvement (Nishimura and Imai 1982).

Endocrine control of renal function may involve a number of factors including both neuro- (e.g. arginine vasotocin, AVT) and

Figure 4.7: Relative proportions (per cent) of 3 types of nephrons (filtering, not filtering and not perfused) in freshwater and sea water adapted rainbow trout (*Salmo gairdneri*) infused with saline (control) or angiotensin II. Each group consisted of 5 trout, each trout providing at least 400 tubules. Means ± standard errors of means shown. Reproduced with permission from Brown *et al.* 1980

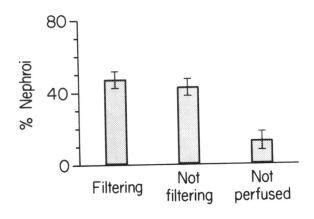

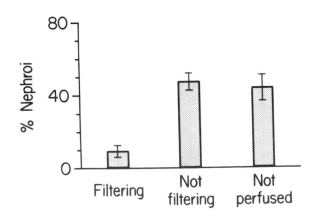

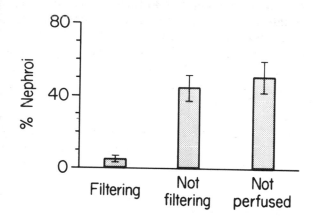

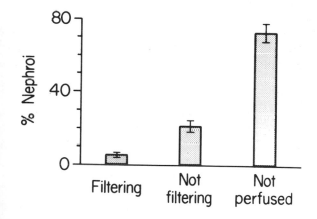

adenohypophysial (e.g. prolactin) hormones, the adrenocortical steroids (the principal one being cortisol), the renin-angiotensin system and catecholamines.

Intravenous injection of AVT causes dose-dependent glomerular diuresis or antidiuresis in the eel (*Anguilla anguilla*) (Henderson and Wales 1974; Babiker and Rankin 1978), suggesting that pituitary release of vasotocin may regulate glomerular activity. Infusion of very small amounts produces antidiuresis in eels in freshwater; secretion of such quantities may promote antidiuresis in sea water-adapted fish. The diuresis provoked by injection of large doses is probably not of physiological significance since the amounts needed are equal to or greater than the total pituitary content. The fact that different doses of the same hormone produce opposite effects may however tell us something about the mechanisms involved in the control of GFR. Pressure in the glomerular capillaries could be varied in several ways. It might be dependent on dorsal aortic pressure, as in the lamprey, but the presence of distinct sphincters in the afferent glomerular arterioles of teleosts (Elger and Hentschel 1981) suggests the possibility of regulation within the kidney. Constriction of the afferent glomerular arterioles would reduce effective filtration pressure as would dilation of the efferent arterioles and vice versa. Vasoactive hormones could produce opposite effects on GFR depending on the sensitivity of receptors mediating vasoconstriction (or vasodilation) in the afferent and efferent arterioles.

Prolactin is especially important for adaptation of teleosts to freshwater (Hirano and Mayer-Gostan 1978). Histological and physiological data show increased prolactin secretion in hypoosmotic environments, although recent work suggests that reduced calcium concentration may be the most important trigger. Its primary effects seem to be to reduce ion and water permeability in the kidney, urinary bladder and gill. A number of histological studies have shown that prolactin affects the structure of the renal tubule cells in both freshwater and sea water adapted teleosts (Olivereau 1980; Wendelaar-Bonga 1976). The actions of prolactin seem to some extent dependent upon cortisol levels. Thus, although cortisol appears to be principally concerned with sea water adaptation it is important in salt conservation of freshwater teleosts where synergistic action of prolactin and cortisol activates sodium pumps in urinary bladder, kidney and gill. Cortisol's effects on membrane permeability are however antagonistic to those of prolactin. It markedly affects enzyme activities of osmoregulatory surfaces, in particular that of sodium + potassium-activated adenosine triphosphatases. Euryhaline fish transferred from fresh to sea water show a progressive increase in the activity of this enzyme in the gills and gut, alongside a reduced renal enzyme activity.

The secretory activity of interrenal tissue (the adrenocortical homologue) is governed by pituitary ACTH and also it seems by the renin angiotensin system (Henderson *et al.* 1976). In euryhaline fish, plasma renin activity may be higher in sea water adapted than in freshwater adapted fish, though this is equivocal (Nishimura 1978). This seems contrary to the position in mammals, where excess sodium inhibits renin release (Davis and Freeman 1976). However, renin release from perfused rainbow trout kidneys was not altered by changes in perfusate sodium concentration; it was increased by reduced perfusion pressure (and also following haemorrhage *in vivo*) (Bailey and Randall 1981). In any case increased renin activity is compatible with the incipient dehydration, increased adrenocortical activity and elevated drinking rates observed in marine teleosts; in most vertebrates angiotensin II (the physiologically-active end product of the renin-angiotensin system) is produced in response to hypovolaemia, stimulates adrenocortical steroidogenesis and promotes drinking.

Angiotensin infusion of freshwater adapted trout induces a glomerular antidiuresis and results in a pattern of renal and glomerular perfusion almost identical to that of sea water adapted trout. Thus angiotensin reduces overall renal arterial blood flow rate (Brown and Oliver 1982) and increases the proportion of non-perfused glomeruli (Brown, Oliver, Henderson and Jackson 1980). Angiotensin infusion into freshwater adapted trout also results in a reduction in the number of filtering glomeruli (Figure 4.7). This may reflect the known vascular actions of angiotensin but may involve a direct effect on the ultrastructure of the glomerular filtration barrier (Brown and Oliver 1982).

Dorsal aortic blood pressure may be an important determinant of effective filtration pressure, and hence SNGFR, and circulating catecholamines play an important role in the control of this parameter (Wahlqvist and Nilsson 1977, 1980) although neural factors may also be involved (Smith 1978). Catecholamines are diuretic in the eel but antidiuretic in the cod (Rankin and Babiker 1981), the nature of the renal response presumably depending on the extent to which pressor effects are compensated for by alterations in renal vascular resistance. Angiotensin II increases dorsal aortic blood pressure, GFR and urine flow in freshwater eels (Nishimura and Sawyer 1976). Part of the pressor response may be ascribed to stimulation of catecholamine release (Nishimura, Norton and Bumpus 1978).

A number of factors may thus be involved in regulation of teleostean renal function but so far the study of all of them is at a preliminary stage and no attempt has been made to integrate them into a general picture.

Kidney Function in Elasmobranchs

Characteristically the elasmobranch nephron is very long, as much as 8 cm compared to a few millimetres in most teleosts, and is made up of several histologically distinct segments. There are, however, far fewer nephrons than in teleosts, for example an 844 g male dogfish (*Scyliorhinus canicula*) had 2280 (Green and Brown, unpubl. obs.) compared to 40 000 in a 200 g trout (Henderson *et al.* 1978). The glomeruli are extremely large and well developed, with many capillary loops, but there is, as yet, no information on their individual rates of filtration. The kidneys excrete small volumes of urine iso-osmotic with the plasma, but with very low levels of urea allowing the high plasma urea levels essential for osmoregulation in these fish to be maintained. Intense urea reabsorption may result from passive movements (Boylan 1972) but a sodium-linked active mechanism has been suggested (Schmidt-Nielsen, Truniger and Rabinowitz 1972) and urea reabsorption can be inhibited by phloretin and chromate ions (Hays, Levin, Meyers, Heinemann, Kaplan, Franki and Berliner 1976). Renal tubular micropuncture is possible in several elasmobranch species and has shown that the nephron establishes large tubular fluid: plasma gradients for several ions, including phosphate, sulphate, calcium and magnesium (Henderson, Brown, Oliver and Hayward 1978).

In contrast to teleosts, there is evidence that dogfish adjust urine production more by alterations in tubular permeability than by changes in GFR. This suggests the possibility of regulation by some antidiuretic factor but little is known about hormonal control of elasmobranch kidney function.

Kidney Function in Lungfish

Although they can, and in the case of the Lepidosirenidae are obliged to, breathe air, lungfish are confined to freshwater except when it ceases to be available during dry seasons. Then African and South

American (but not Australian) species aestivate in cocoons in the mud, and urine production ceases completely (Delaney, Lahiri, Hamilton and Fishman 1977). They may spend several years in this condition, gradually losing water by evaporation from the lungs and accumulating urea from protein catabolism until plasma osmolarity is greatly increased. On re-immersion they rapidly absorb water, swelling considerably and becoming very diuretic.

Control of renal function in lungfish has recently been reviewed (Sawyer, Uchiyama and Pang 1982) so only a few points will be made here. Changes in urine flow result from changes in GFR, and hormones which increase dorsal aortic blood pressure, such as AVT and catecholamines are usually, but not always, diuretic. In the cases where they are not, direct action on the renal vasculature must compensate for the pressor effect, In the Australian lungfish (*Neoceratodus*) AVT increases dorsal aortic pressure and GFR synchronously (Sawyer, Blair-West, Simpson and Sawyer 1976) but in the African lungfish (*Protopterus*) the diuresis may outlast the pressor response (Sawyer 1970; Babiker and Rankin 1979b). This could be due to an action of AVT on glomerular arterioles or to renal vessels taking some time to return to their resting state following dilation. Lungfish renal function thus seems to be largely controlled by the balance between dorsal aortic blood pressure and the tone of the renal vasculature. No antidiuretic actions of neurohypophysial hormones have been demonstrated.

Summary

Our knowledge of the control of fish renal physiology is clearly still fragmentary. Nothing is known about the possibility of nervous control and although a number of hormones have been implicated only a few have been studied in any detail, usually in isolation. Only some of the variety of hormones which can affect renal function have been mentioned above; the roles of others, such as parathyroid hormone, which is antidiuretic in lungfish (Pang and Sawyer 1975), are even more obscure.

Urine flow rate in freshwater fish seems to depend on GFR, tubular fluid reabsoption being low and tending not to vary (although it is very dangerous to generalise about such a diverse group, so few representatives of which have been studied). GFR depends on glomerular capillary blood pressure, and possibly flow. Changes in dorsal aortic blood pressure may be directly reflected in changes in GFR, or control of the renal vasculature may make independent control of GFR possible. Progress will only come from a greater understanding of the control of the renal circulation.

In marine fish tubular secretion of ions and fluid is of greater (or in the case of aglomerular species, sole) importance, but the few studies on its hormonal control have failed to reveal a clear picture (Babiker and Rankin 1979a; Churchill, Malvin, Churchill and McDonald 1979; Zucker and Nishimura 1981).

Acknowledgement

The authors' work on the subject was supported by the Science and Engineering Research Council and the National Environment Research Council.

5   PANCREATIC CONTROL OF METABOLISM

Bernard W. Ince

'Regulation of nutrient homeostasis constitutes one of the most important of all physiological functions, determining the ability of organisms to perform optimally and thereby survive in an environment that imposes unpredictable changes in the supply of and demand for fuels ... charged with the formidable task of regulating nutrient homeostasis, are the islets of Langerhans' (Unger, Dobbs and Orci 1978)

Bearing in mind the mammalian context of this statement, what evidence is there in support of such a role for the pancreatic islets of teleost fish, the most diverse of all vertebrates, surviving as they do in unpredictable aquatic environments? In order to provide some answers to this question, the current state of knowledge regarding pancreatic control of metabolism (and other processes), and of the factors which influence islet activity in teleosts, are presented. Moreover, whilst those aspects which need more study are clearly revealed, attention is also focused on the use of alternative experimental approaches which might possibly provide more penetrating information. A theme of this nature does not warrant an extensive literature coverage. Accordingly, the references given relate primarily to studies conducted since this field was last reviewed in depth (Epple 1969; Brinn 1973). The more detailed consideration of insulin's role, apparent throughout, merely reflects what little is known by comparison for the other islet hormones.

Role of Insulin

Carbohydrate Metabolism

The hypoglycaemic response to exogenous insulin is probably the single most recorded observation of insulin action in teleosts (Epple 1969; Ince and Thorpe 1974; Lewander, Dave, Johansson-Sjobeck, Larsson and Lidman 1976; Akhtar, Butt and Ahmad, 1979; Ablett, Sinnhuber, Holmes and Selivonchick 1981). Hypoglycaemia is characteristically prolonged, especially at low temperatures, but of shorter duration at higher temperature (Murat and Serfaty 1975). In any one species, however, both magnitude and duration are dependent on a number of other factors, including insulin type and dose level, route of injection, season, nutritional state, and previous nutritional history. Interspecific variations in response to insulin may relate partly to differences in mode of life (Leibson 1972; Thorpe and Ince 1976). The resistance of teleosts to pharmacologic doses of insulin, resulting in low or negligible blood glucose levels, is well known (Epple 1969). In mammals, perhaps the most important role of the islet hormones is the regulation of glucose homeostasis, in order to maintain the functional integrity of the brain and central nervous system. Evidently, this requirement is of lower priority in teleosts, perhaps partly on account of their higher brain glycogen reserves (Plisetskaya 1968).

The effects of insulin on glycogen synthesis in various tissues have also been studied *in vivo* and *in vitro*. Liver glycogen levels, however, often decrease *in vivo* following insulin injection, whereas muscle glycogen is less affected (Lewander *et al.* 1976; Ince and Thorpe 1976; Ablett, Sinnhuber and Selivonchick 1981). A more consistent glycogenic response to insulin has been found *in vitro* (Tashima and Cahill 1968; deVlaming and Pardo 1975). The paradoxical response *in vivo* may therefore result from the action of counter-regulatory hormones, released in response to hypoglycaemia, although this possibility has not been studied. Similarly, the mechanisms underlying the hypoglycaemic response to insulin have received insufficient study. However, these most likely relate to a combination of increased oxidative clearance of glucose in peripheral tissues (Ablett, Sinnhuber and Selivonchick 1981; Ablett *et al.* 1981), and reduced hepatic glucose production (Hayashi and Ooshiro 1975). The latter authors made use of an isolated, perfused eel liver preparation. Techniques of this kind, rather than the more common single-injection approach, are likely to prove of greater value in the future, when investigating the interrelated metabolic actions of islet hormones.

Overall, the evidence suggests that insulin's role in teleost carbo-hydrate metabolism is directed more towards oxidative clearance of glucose than its deposition as glycogen.

Protein Metabolism

It is now generally accepted that the major role of insulin in teleosts is the regulation of protein metabolism (Thorpe 1976; Murat, Plisetskaya and Woo 1981). Insulin decreases plasma amino nitrogen (Palmer and Ryman 1972; Thorpe and Ince 1974; Ince and Thorpe 1974), individual amino acids (Inui, Arai and Yokote 1975; Cowey, De Lattiguera and Adron 1977; Ince and Thorpe 1978a), and protein turnover *in vivo* (Castilla and Murat 1975); increases the incorporation of ^{14}C-labelled amino acids into liver and/or skeletal muscle protein *in vivo* and *in vitro* (Tashima and Cahill 1968; Jackim and La Roche 1973; Ahmad and Matty 1975a; Ince and Thorpe 1976; Ablett, Sinnhuber and Selivonchick 1981; Ablett *et al.* 1981), and suppresses gluconeogenesis from amino acids *in vivo* (Inui *et al.* 1975; Cowey *et al.* 1977). Insulin thus enhances proteogenesis, whilst suppressing gluconeogenesis, and exerts these actions in both fed and fasted fish (Cowey *et al.* 1977; Ablett *et al.* 1981).

Lipid Metabolism

Evidence that insulin enhances lipogenic pathways in teleosts has also been provided in recent studies. A reduction in plasma fatty acids and cholesterol *in vivo* (Minick and Chavin 1972; Palmer and Ryman 1971; Ince and Thorpe 1975; Lewander *et al.* 1976), and increased incorporation of carbon derived from glucose and amino acids into liver and/or skeletal muscle lipid *in vivo* and *in vitro* (Tashima and Cahill 1968; deVlaming and Pardo 1975; Ince and Thorpe 1976; Ablett, Sinnhuber and Selivonchick 1981), have been observed following insulin treatment. In fasted fish, insulin exerts a preferential effect on lipogenesis from glucose in skeletal muscle *in vivo*, which taken with its proteogenic action on this tissue, suggests that the hormone counteracts excessive depletion of these fuels during periods of food deprivation. The mechanisms whereby insulin exerts its antilipolytic actions, however, may be different from those in mammals (Plisetskaya 1980).

The data presented in Figure 5.1 exemplify the wide-ranging metabolic actions of insulin in teleosts. The greater efficacy of fish as compared with mammalian insulin, is also apparent (Ince and Thorpe 1974).

Growth

If insulin enhances protein anabolic processes in teleosts, can it justifiably be considered a growth-promoting hormone? Few studies have attempted to answer this important question. Ludwig, Higgs, Fagerlund and McBride (1977) were first to study the possible

Figure 5.1: Effects of a single intravascular injection of bovine insulin (open squares) and codfish insulin (open triangles) at 2 IU/kg body weight on (a) blood glucose, (b) plasma cholesterol, and (c) plasma amino acid nitrogen, in European silver eels (*Anguilla anguilla*). Each fish was cannulated to permit serial blood sampling. Filled circles represent control animals. Values are means ± SE, with the number of animals in parenthesis. Codfish insulin causes a progressive lowering of blood glucose, plasma cholesterol and amino acid nitrogen, whereas bovine insulin is effective in lowering only amino acid nitrogen. Recovery of blood metabolites occurs 24-48 h after insulin injection.

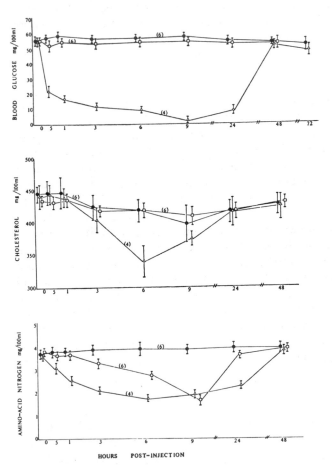

Reproduced by permission of *Gen. Comp. Endocrinol.*, from Ince and Thorpe 1974

involvement of insulin in teleost growth. Prolonged administration of bovine insulin to coho salmon (*Oncorhynchus kisutch*) increased specific growth rate, and decreased food-gain ratios relative to solvent-injected controls, although the effects were not statistically significant. Using a high bovine insulin dose range (0.5-5.0 IU/kg), Ablett, Sinnhuber and Selivonchick (1981) observed a significant increase (18.5 per cent) in body weight of rainbow trout (*Salmo gairdneri*) after 56 days, together with increases in liver-somatic indices and condition factor. It is not known, however, to what extent insulin may influence food intake, or how cellular growth (in terms of RNA and DNA) in different tissues responds to the hormone. In mammals, growth hormone is less effective in the absence of insulin (Daughaday, Herington and Phillips 1975), a further aspect worthy of study in teleosts. The use of fish, as opposed to mammalian, insulin, would clearly be preferable in studies of this nature.

Hydromineral Balance

Pancreatic hormones have also been implicated in teleost hydro-mineral balance, although such studies have been performed only in eels. Epple and Lewis (1975) observed higher mortalities of pan-createctomised eels (*Anguilla rostrata*) in seawater than freshwater. Pancreatectomy was accompanied by tissue hydration, and when combined with hypophysectomy ('Houssay operation'), eels could only survive in 1/3 seawater (Epple and Lewis 1977). More recent studies in eels have indicated that islet hormones may contribute to the regulation of intracellular free amino acid levels in hyperosmotic media (Epple and Koscis 1980), although whether insulin alone is responsible remains to be clarified. Evidence that one possible site of insulin action may be the gills has been provided by Scheer and Langford (1976). Thus, although the hormone was without effect on gill ATPase activity, total Na efflux was increased, whilst the K-dependent fraction was reduced, suggesting that insulin may offset Na influx after seawater transfer of eels (*A. anguilla*). The effect of insulin on renal electrolyte metabolism in teleosts, however, appears not to have been studied.

Role of Glucagon

Exogenous administration of glucagon, elicits hyperglycaemia in teleosts (Epple 1969; Larsson and Lewander 1972; Thorpe and Ince

1974; Bhatt and Khanna 1976; Ince and Thorpe 1977a; Chan and Woo 1978), as a result of glycogenolysis, primarily in the liver (Umminger, Benziger and Levy 1975; Birnbaum, Schulz and Fain 1976). Glucagon has also recently been shown to enhance gluconeogenesis in a variety of species (eel, trout, carp), and thus opposes the proteogenic actions of insulin (Inui and Yokote 1977; Castilla and Murat 1975; Walton and Cowey 1979). Chan and Woo (1978), in a detailed metabolic balance study, found that glucagon, whilst increasing net protein and carbohydrate catabolism, was without effect on fat oxidation in eels (*A. japonica*). In certain species (toadfish, eel), glucagon's inability to increase plasma fatty acids (Plisetskaya 1980), except at very high dose levels (Ince and Thorpe 1975), suggests that the hormone does not oppose the antilipolytic actions of insulin. However, more studies are required to define glucagon's role in teleost lipid metabolism.

Regulation of Insulin Release

The most significant advance made in the last decade has been the development of fish insulin radioimmunoassay systems (Patent and Foa 1971; Plisetskaya, Leibush and Bondareva 1976; Thorpe and Ince 1976; Furuichi, Nakamura and Yone 1980). Much valuable information on the factors regulating B-cell secretory activity has recently been obtained using this technique, but little is known for either glucagon or somatostatin at the present time.

Nutrients and Nutritional State

In man, glucose is the major stimulus for insulin release and blood glucose regulation is probably the single most important function of the pancreatic islets (Unger *et al.* 1978). The situation in teleosts, however, is somewhat different. Their higher dietary protein requirements for optimal growth than mammals, and highly efficient gluconeogenic capacity, contrasts with a less efficient utilisation of dietary carbohydrate (Cowey and Sargent 1979). The inferior tolerance to a glucose load by comparison with an amino acid load, for example, is well known (Palmer and Ryman 1972; Thorpe and Ince 1974). Moreover, a positive correlation exists between dietary protein level and plasma insulin (Ahmad and Matty 1975b), and an inverse correlation with respect to dietary carbohydrate (Yone 1978).

This situation is also reflected in the differential sensitivity of teleost B-cells to individual nutrients *in vivo* and *in vitro*. In eels, for example, lysine and arginine are potent insulin secretagogues by comparison with glucose (Ince and Thorpe 1977b; Ince 1979), although not all amino acids are as effective (Patent and Foa 1971), nor is glucose necessarily the most insulinogenic of hexoses (Fletcher, Noe and Hunt 1978). Thus, considerable variations are likely to exist depending on the specific nutritional requirements of different species. The differential secretory response of glucagon and somatostatin to glucose and amino acids has been studied in only a single instance to date. Ronner and Scarpa (1981) showed arginine to be a potent stimulator of glucagon release from perfused channel catfish Brockmann bodies *in vitro*, but neither this amino acid nor leucine stimulated somatostatin release. Addtionally, whereas glucose (10 mM) stimulated somatostatin release, the release of glucagon was only elicited at low glucose (0.2 mM) concentrations (as in mammals), implying that high glucose levels suppress glucagon release. The isolation of teleost pancreatic islets using collagenase, and their maintenance in culture media, has also recently been shown to be technically feasible by Thorpe and Duve (1980). The potential of such techniques in the present context is evident, but remains to be fully exploited.

Changes in nutritional state are also accompanied by corresponding changes in plasma insulin levels (Tashima and Cahill 1968; Patent and Foa 1971; Thorpe and Ince 1976). Insulin levels are higher in fed than in fasted fish, and appear to correlate more closely with plasma amino nitrogen than glucose, the latter showing little variation (Table 5.1). During fasting, blood glucose levels are maintained through gluconeogenesis (Cowey and Sargent 1979). Insulin may therefore exert some indirect control over blood glucose by regulating the extent of skeletal muscle catabolism, and/or by enhancing metabolic rate and peripheral glucose utilisation (Ablett, Sinnhuber and Selivonchick 1981; Ablett *et al.* 1981). A reduced endogenous delivery of insulin would also tend to favour increased hepatic glucose output. However, what mechanisms operate to maintain endogenous insulin delivery during fasting, and further, what influence gastrointestinal hormones and neural stimuli contribute to insulin release during the digestive phase, are questions which need answering. The role of fatty acids in regulating insulin release in teleosts, likewise, remains to be investigated.

Table 5.1: Effects of starvation and feeding in rainbow trout. Each group serially sampled three times. Values are means ± SE, with number of animals/group in parenthesis.

	Weight (g)	Plasma insulin (ng/ml)	Plasma glucose (mg/dl)	Plasma amino acid nitrogen (mg/dl)
Group 1 ($n = 10$)				
Starved 7 days	216 ± 7	1.60 ± 0.30	77.5 ± 8.3	9.61 ± 0.37
Fed 7 days	232 ± 8	6.80 ± 0.60*	80.0 ± 6.8	12.10 ± 0.80*
Starved 7 days	223 ± 6	2.30 ± 0.40*	73.5 ± 2.8	7.00 ± 0.71*
Group 2 ($n = 10$)				
Fed 7 days	231 ± 7	4.21 ± 0.32	77.8 ± 6.8	12.21 ± 0.92
Fed 14 days	227 ± 7	5.65 ± 0.30	66.4 ± 8.0	11.55 ± 0.88
Starved 7 days	215 ± 6	2.20 ± 0.30*	63.5 ± 3.4	7.76 ± 0.76

Note: * $P < 0.05$ between treatments within each group.

Source: Reproduced by permission of *General and Comparative Endocrinology*, from Thorpe and Ince 1976.

Seasonal Variations

Plasma insulin levels in teleosts show marked seasonal changes in relation to gonad maturation and reproductive activity (Plisetskaya, Leibush and Bondareva 1976; Singley and Chavin 1976; Plisetskaya, Soltitskaya and Leibson 1977). In the non-migratory scorpionfish (*Scorpaena porcus*) for example, a fall in plasma insulin occurs during the winter months, although complete fasting does not take place. During spawning and pre-spawning periods, feeding continues (in contrast to most migratory species), and insulin levels are elevated. The cause of these changes is not known with certainty, however, but most likely result from a combination of environmental and endogenous signals. Spawning is governed primarily by water temperature, and at this time, islet hypertrophy and improved glucose and amino acid tolerance have been observed (Palmer and Ryman 1972; Plisetskaya, Leibush and Bondareva 1976). The functional significance of hyperinsulinaemia during these periods, is equally unclear, but it is nevertheless tempting to speculate that insulin may be involved in some aspect of gonad maturation and/or spawning activity.

Hormonal Influences

Many hormones are known to exert either direct or indirect influences on islet physiology in mammals (Gerich, Charles and

Grodsky 1976). In teleosts, studies of this nature have only recently been conducted, the majority with regard to the catecholamine hormones, adrenaline and noradrenaline. These compounds, acting either as hormones or as neurotransmitters, exert a dual effect on insulin release in mammals through activation of an α-adrenergic (suppression) and β-adrenergic (stimulation) B-cell receptor (Smith and Porter 1976). In teleosts, almost identical responses have been observed, as illustrated in Figure 5.2, suggesting that a similar control mechanism is operative (Plisetskaya *et al.* 1976; Ince and Thorpe 1977a; Ince 1980). It has been suggested that during periods of high activity, when circulating catecholamine levels are elevated, reduced output of endogenous insulin would be of adaptatory value in permitting a more rapid mobilisation of metabolic fuels (Ince and Thorpe 1977a). This hypothesis, however, has yet to be rigorously tested.

The influence of other hormones has received less study. Mammalian glucagon is without effect on plasma insulin levels in eels, despite stimulating hyperglycaemia (Ince and Thorpe 1977a), but elicits insulin release in scorpionfish (Plisetskaya *et al.* 1976). A secretin-like factor, extracted from pike intestines, has recently been found to stimulate insulin release *in vivo* in eels (Ince 1983), suggesting that a functional relationship between the gut and islets may exist in teleosts. However, perhaps the most significant finding is that somatostatin inhibits insulin release in teleosts (Ince 1980; Oyama *et al.* 1981), as it does in mammals (Unger *et al.* 1978). Although not exclusively localised in the islet D-cells of teleosts, pancreatic somatostatin would nevertheless be strategically placed to exert direct regulatory influences on B-cell activity, and thus modulate insulin-mediated metabolic events.

Mechanisms of Insulin Release

A vast literature has accumulated on the mechanisms of insulin release in higher vertebrates, directed primarily towards a better understanding of abnormal B-cell function in diabetes mellitus (Gerich *et al.* 1976). However, whilst comparative studies in teleosts could be valuable in this context, the available information is very fragmentary, despite the unique nature of their Brockmann bodies, an ideal experimental tissue for such investigations. Recently, however, an *in situ* perfused pancreas technique for the eel (*A. anguilla*) was developed as an alternative for studying mechanistic and dynamic aspects of secretion of islet hormones (Ince 1979). A

Figure 5.2: Effects of a single intravascular injection of adrenaline on plasma insulin and glucose levels in cannulated European silver eels (*Anguilla anguilla*). Values are means ± SE, with the number of animals in parenthesis. Circles and squares represent dose levels of 25 and 50 µg/kg body weight respectively. At both dose levels, a biphasic response of plasma insulin occurs, an initial depression after 30 min., corresponding to the evolution of hyperglycaemia, followed by an elevation ('rebound effect'), corresponding to recovery from hyperglycaemia. Similar responses to noradrenaline also occur.

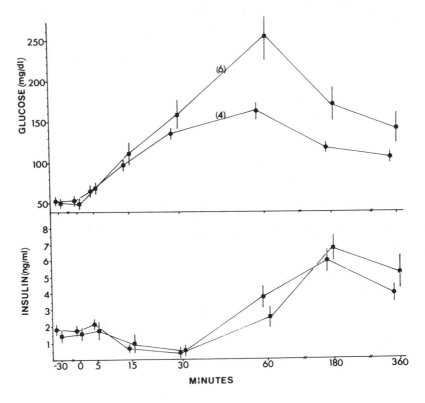

Reproduced by permission of *Gen. Comp. Endocrin.*, from Ince and Thorpe 1977a

schematic representation of the very simple perfusion system needed is illustrated in Figure 5.3. Using this technique, it has been shown that insulin secretion in response to a constant glucose or amino acid stimulus, is biphasic, in accord with observations in other vertebrates (Figure 5.4). The intracellular events mediating release are, nevertheless, poorly understood. It is known, however, that teleost islets discriminate between D- and L-glucose, and that the D isomer equilibrates rapidly across the islet cell membrane (Cooperstein and Lazarow 1969). Glucose transport *per se* is therefore unlikely to be rate-limiting for insulin release. However, if the rate of glucose metabolism in the islets is any reflection of the situation in extrapancreatic tissues, then its phosphorylation to glucose-6-phosphate may partly determine the B-cell's capacity for responding to a glucose stimulus. Support for this view comes from the studies of Watkins (1972), who observed that insulin release from toadfish islets *in vitro* was almost identical in the presence or absence of glucose (300 mg/dl), but was significantly enhanced in the presence of reduced pyridine nucleotides (NADH, NADPH), but not their oxidised forms. Glucose may therefore be more important to insulin biosynthesis than to release (Fletcher *et al.* 1978). Evidence that cAMP is involved in insulin release from teleost islets has been obtained using the perfused eel pancreas. Theophylline, a phosphodiesterase inhibitor, which increases intracellular cAMP levels, stimulates biphasic insulin release in the presence of substimulatory glucose levels (2.7 mM), and potentiates lysine-stimulated insulin release (Ince 1980). It remains to be determined, however, whether cAMP acts in a modulatory capacity, or more directly on the secretory process itself.

Insulin Dynamics

The *in vivo* dynamics and metabolism of insulin have received little study in teleosts, at least by comparison with the thyroid and steroid hormones, and may appear to be a rather peripheral issue in the context of the preceding discussion. Nevertheless a close relationship between receptor binding and hormone metabolism has been identified in mammals, which may have an important bearing on the physiological actions of insulin, and the maintenance of its steady-state plasma levels (Sönksen, Jones, Tompkins, Srivastava and Nabarro 1976). The marked seasonal changes in plasma insulin, for example, must partly reflect changes in the balance between

Figure 5.3: Schematic representation of the *in situ* perfused eel pancreas system. The control and experimental media (eel Ringer containing 2 per cent dextran 40, 0.3 per cent albumin and 2.7 mM glucose), are gassed continuously with 95 per cent O_2: 5 per cent CO_2, and pumped through a glass coil placed in a thermostatically controlled water bath to maintain perfusate temperature at 21°C. The perfusate selector mechanism (three-way taps) allows either independent recirculation of both media, or redirection of perfusate into the organ preparation through the arterial cannula. A pressure transducer linked to a chart recorder monitors perfusion pressure. During surgical isolation of the pancreas, anoxia is minimised by vigorous irrigation of the gills with aerated saline. Perfusate samples are collected into chilled, graduated vessels, and stored at — 20°C, prior to hormone assay.

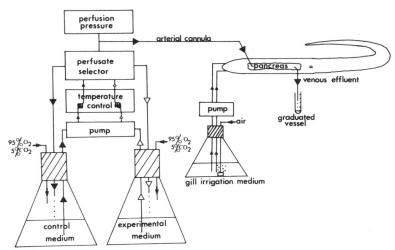

Reproduced by permission of *Gen. Comp. Endocrin.*, from Ince 1979

endogenous insulin production and elimination, the kinetic characteristics of which deserve study *in vivo*, and using individual organ systems. The few investigations performed to date, have been largely methodological, exploring various experimental approaches, and kinetic methods of analysis, so that the most appropriate conditions for studying insulin dynamics in teleosts may be defined (Ince and Thorpe 1978b; Ince 1982). Preliminary results indicate that the kidneys and liver (and perhaps, the gills) of teleosts, as in mammals, may play a key role in the dynamics of insulin.

Figure 5.4: Basal insulin secretion and secretory response of the perfused eel pancreas to 30 mM D-glucose (triangles) and to 10 mM (squares) and 20 mM (circles) L-arginine hydrochloride. Values are means ± SE, with the number of animals in parenthesis. A pre-stimulatory period (—15-0 min) of basal insulin secretion in the presence of 2.7 mM glucose, is followed by a stimulatory period of 60 min (indicated by vertical arrows), and a post-stimulatory period (60-80 min) in the presence of 2.7 mM glucose. A biphasic pattern of insulin secretion to 30 mM glucose and to arginine is evident, the latter evoking a significantly greater response.

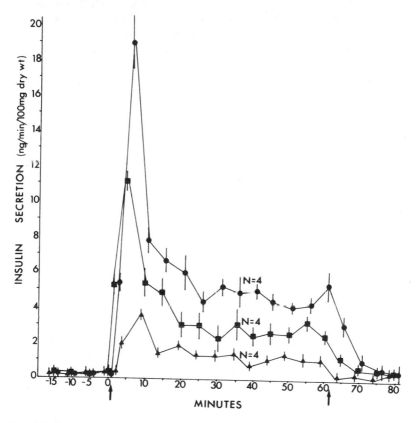

Reproduced by permission of *Gen. Comp. Endocrinol.*, from Ince 1979

Summary

(1)   Insulin enhances proteogenesis and lipogenesis in teleosts, whilst its role in carbohydrate metabolism seems to be directed more towards glucose oxidation than glycogenesis. Growth and osmo-regulation are two further processes in which insulin and/or islet hormones have been implicated, but studies in more species are needed for confirmation.

(2)   Glucagen stimulates glycogenolysis and gluconeogenesis, but there is no convincing evidence or a lipolytic role in teleosts.

(3)   Plasma insulin levels are dependent on dietary protein level, nutritional state, and season. Amino acids provide a greater stimulus for insulin release than glucose, although the intracellular mechanisms involved are poorly understood. Circulating catecholamines both inhibit and stimulate insulin release. Somatostatin exerts a strong inhibitory influence.

(4)   The effects of fatty acids, gastrointestinal hormones and neural influences on insulin release, and of altered hormone dynamics on steady-rate insulin levels, are unknown, and warrant immediate study.

(5)   The factors regulating glucagon and somatostatin levels and release in teleosts, remain to be determined. The acquisition of such information is arguably of the highest priority at the present time.

6 SYNTHETIC CHEMICALS AND PHEROMONES IN HOMING SALMON

Arthur D. Hasler

The olfactory hypothesis for salmon homing (Hasler and Wisby 1951) states that (a) before juvenile salmon migrate to the sea they become imprinted to the distinctive odour of their natal tributary; and (b) adult salmon use this information as a cue for homing when they migrate through the homestream network to the home tributary.

Restating this in more precise terms, the use of olfaction for homing by salmon requires that:

(1) Because of its unique vegetation and soil types, each stream must have a characteristic and persistent odour perceptible by the salmon.

(2) They must be able to discriminate between the odours of different streams.

(3) Salmonids must be able to retain an 'odour memory' of the homestream during the period that intervenes between the downstream migration and homing migration.

In this chapter I review the behavioural and physiological experiments conducted in our laboratory that support these three postulates.

Conditioning and Sensory Impairment Experiments

The first step in testing the odour hypothesis was to determine if fish could discriminate between two different streams by smell (Hasler and Wisby 1951). Using reward (food) and punishment (electric shock) for conditioning, we trained groups of coho salmon to discriminate between waters collected from two streams. However, when their nasal sacs were cauterised trained fish were not able to discriminate these waters. This study implied that each stream has a characteristic odour that is discernible by fish.

This study also provided information on two additional points. First, trained fish were not able to identify a water sample if the organic fraction was removed. Secondly, fish that were trained to

103

discriminate water obtained from a stream during one season were able to discriminate water from the same stream collected in a different season. This result suggests that the factor that the fish could detect was long-lasting and present in the stream throughout the year. This, we believed, was an important point because salmon may reside in the ocean for several years before returning to their natal stream — thus the imprinting factor must be long-lasting. This experiment was repeated by other investigators (Idler, Bitners and Schmidt 1961; McBride, Fagerlund, Smith and Tomlinson 1964; Tarrant 1966; Walker 1967) whose results were similar to ours.

Our second step in testing the odour hypothesis was to conduct sensory impairment experiments in the field (Wisby and Hasler 1954). The purpose of these experiments was to determine if fish deprived of their sense of smell could locate their homestream. The research site was a small Y-shaped tributary located 25 km from Seattle, Washington: Issaquah Creek and its East Fork (Figure 6.1). Each branch had its own native stock of coho salmon. We captured 322 fish in traps as they entered their home branches and plugged the nasal sacs in half of them to block their olfactory sense. The remaining fish were left unplugged to control for olfactory impairment. We also tagged each fish to tell us which treatment it had received and in which branch it had been caught. All of the fish were then released 1.6 km below the fork and allowed to repeat their upstream migration. The control fish returned to their home tributary almost without error, but many of the fish that had been deprived of their sense of smell entered the wrong branch of the stream. The results are summarised in Table 6.1.

This type of sensory impairment experiment has been repeated 20 times by other investigators (reviewed by Hasley 1966; Stasko 1971). Species tested include coho (*Oncorhynchus kisutch*), chum (*O. keta*), chinook (*O. tschawytscha*), and Atlantic salmon (*Salmo salar*), and cutthroat (*S. clarki*), rainbow (*S. gairdneri*), and brown trout (*S. trutta*). The results of these studies are remarkably consistent and agree closely with the results of our own experiment. For 16 experiments the olfactory sense appeared to be necessary for correct homing. In addition, two studies demonstrated that blinded fish homed nearly as well as control fish, and thus that vision was not so important for relocating the original stream, at least during the upstream migration (Hiyama, Taniuchi, Suyama, Iashoka, Sato and Kajihara 1966; Groves, Collins and Trefetren 1968).

Up to this point our experiments showed that each stream has a

Figure 6.1: Study area for olfactory impairment experiments in Lake Washington watershed, adapted from Wisby and Hasler (1954). Inset shows detail of study site at Issaquah Creek and its East Fork.

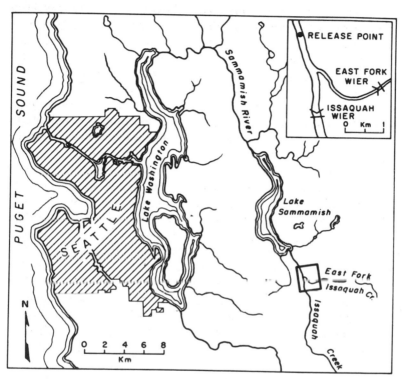

Table 6.1: Results of nose-plugging experiments on coho salmon conducted at Issaquah Creek in 1954

Stream of origin	Treatment of fish	Number released	Number recaptured in	
			Issaquah	East Fork
Issaquah	Controls	121	46	0
	Nose plugged	125	39	12
East Fork	Controls	38	8	19
	Nose plugged	38	16	3

Source: from Wisby and Hasler 1954

characteristic and persistent odour, and that salmon can discriminate between the odours of different streams, thus supporting the first two tenets of the olfactory hypothesis. However, these studies did not provide conclusive evidence for olfactory imprinting and long-term olfactory memory, hence we modified our plan of research to test this point of the hypothesis.

Evidence for Imprinting — Transplantation Experiments

Central to our basic design were the results obtained from experiments with fish transplanted from their natal tributary to a different tributary during the smolt stage, i.e. at the time they begin to make their seaward journey.

When Rounsefell and Kelez (1938) transferred young, presmolt coho salmon from their native river to a different river where they marked and released them, the fish migrated to sea and returned as adults to the second river. Donaldson and Allen (1957) also found that coho salmon raised in a hatchery and transplanted before undergoing smolt transformation returned to the river of release. There is evidence that this process is rapid. Jensen and Duncan (1971) and Carlin (1968) transplanted coho salmon and Atlantic salmon, respectively, just as they began to smolt; the fish left the river within two days and returned to it during the spawning migration.

In contrast, when Peck (1970) did not transplant hatchery-raised coho salmon in a Lake Superior tributary until several weeks after smolt transformation, the return to the stream of release was poor and many fish were recovered in other streams. Peck concluded that the fish may have spent the sensitive period in the hatchery (water supply not connected with Lake Superior) before stocking and suggested that imprinting terminates soon after smolt transformation begins, thereby preventing fish from being imprinted to other tributaries during their downstream migration. Additional support for this view comes from a study by Stuart (1959) working with brown trout at Dunalastair Reservoir in Scotland. He held fish in their native tributary through the smolt stage before transferring them to a different tributary. During the spawning migration the adults returned to their natal stream. However, fish transplanted before undergoing smolt transformation returned to the river of release.

Thus, results from transplantation experiments suggest that homing is connected with a period of rapid and irreversible learning (or 'imprinting') of the cues that identify the homestream at the time the young salmon begin their downstream migration, i.e. smolt transformation.

Artificial Imprinting Experiments

We reasoned that a decisive test for odour imprinting was to substitute a man-made chemical for the natural scents of the homestream (Hasler and Wisby 1951). We could then expose salmon to the chemical during the smolt stage (the 'critical' or 'sensitive' period for imprinting — see transplant experiments above) and try to decoy them to a stream scented with it during the spawning migration.

We decided that the chemical had to be an organic compound since our previous work had demonstrated that the identifiable component of stream water was contained in the organic fraction. In addition, it would have to be chemically stable, soluble in water, not normally found in natural waters, and one that would neither naturally repel nor attract the fish. Finally, we felt that it was important for the chemical to be detectable in small quantities in order to minimise any possible damage it might have on a natural stream system. Wisby (1952) screened likely substances and selected morpholine (C_4H_9NO), a heterocyclic amine, which could be detected by unconditioned coho salmon at a concentration of 1×10^{-6} mg/l.

A series of experiments was performed by Hirsch (1977) to determine if heart rate changes in coho salmon could be conditioned to odour stimuli. It was concluded that:

(1)   Heart-rate changes were conditioned in coho salmon to chemical stimulation of the olfactory system.

(a) Presentation of the chemical without the unconditional shock stimulus had no apparent or drug-like effect on the fish.

(b) Control fish placed in a duplicate chamber of the apparatus and subjected to the shock stimulus but without the odour stimulus did not condition to accompanying stimuli.

(c) Fish with occluded nares did not become conditioned to the odour associated with shock stimuli; though they did so after the nares were unplugged. Occluding the nares of conditioned fish eliminated the response to the odour.

(2)   Fish conditioned to respond to morpholine did not respond to phenethyl alcohol (PEA), although they could be conditioned to do

so. This suggests that fish were not only able to detect the odour stimulus, but were able to discriminate between two odour stimuli. (3)   Fish could readily detect morpholine at 2×10^{-3} mg/l and did not seem to at 2×10^{-5} mg/l. However due to technical difficulties, this lower limit may need revision.

We started artificially imprinting salmon on a large scale in the spring of 1971 (Madison, Scholz, Cooper, Horrall, Hasler and Dizon 1973; Scholz, Gosse, Cooper, Madison, Horrall and Hasler 1973, Cooper, Scholz, Horrall, Hasler and Madison 1976). We transported 16 000 coho fingerlings of the same genetic stock and hatched and raised under uniform conditions, from a Wisconsin State Fish Hatchery in central Wisconsin to holding tanks at a water-filtration plant in South Milwaukee (Figure 6.2). We held the fish there for about 30 days during their smolting period. All the fish were held in tanks with water piped in from Lake Michigan; thus, during their entire early life history they were not exposed to water from any Lake Michigan tributary.

Morpholine was metered into a tank containing half of the fish. A concentration of 5×10^{-5} mg/l was maintained in the tank throughout the exposure period. The rest of the fish, held in a second tank, were not treated, to serve as a control group. The fish from each group were then marked with identifying fin clips and released into Lake Michigan 0.5 km south of the mouth of Oak Creek, the stream which was to be scented during the spawning migration. We released them directly into the lake in order to reduce the possibility that they might learn additional cues about the test stream.

During the spawning season in the fall of 1972, 18 months after the fish were released, we created an artificial homestream for the homing adults by metering morpholine into Oak Creek at approximately the same concentration to which they had been exposed. We hypothesised that if salmon use a chemical imprinting mechanism for homing, morpholine-exposed fish would return to the scented stream in larger numbers than unexposed fish. Unexposed fish served as controls to determine if fish would return to the stream independently of the chemical cue. The stream was monitored by creel census surveys, gill netting, and electrofishing. We captured a total of 216 of the fish that had been exposed to morpholine and only 27 from the control group. This result supported our hypothesis but we decided to repeat the experiment.

During the spring of 1972, we started experiments to replicate the 1971-72 series (Scholz, Horrall, Cooper and Hasler 1975, Cooper *et*

Figure 6.2: Research area, South Milwaukee, Wisconsin. Inset A shows detail of: (1) the water intake for the ranks at the South Milwaukee water filtration plant, (2) the Oak Creek stocking site, and (3) the Milwaukee Harbor stocking site.

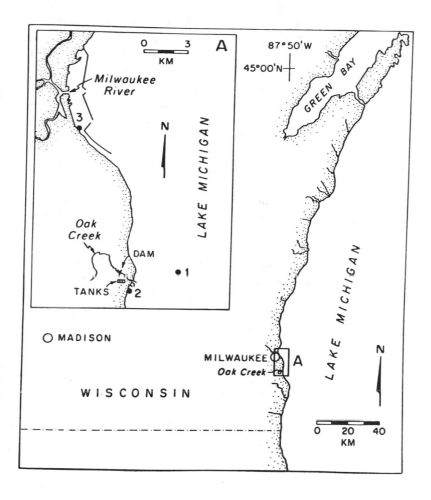

al. 1976). We worked with larger numbers of smolts: 18 200 were exposed to morpholine at a hatchery and 20 000 were left unexposed. This time we released the fish at two different points along the Lake Michigan shoreline, 0.5 km south and 13 km north of Oak Creek (see Figure 6.2). In the fall of 1973, we caught 1515 morpholine-exposed fish and 169 controls — a ratio of nearly 10 to 1 (Table 6.2). Morpholine-exposed fish released as smolts 13 km north of Oak Creek homed to Oak Creek in a manner similar to those released near it (Table 6.2).

During a third control experiment conducted in 1973, we exposed 5000 smolts to morpholine and left an equal number unexposed (Scholz *et al.* 1975; Cooper *et al.* 1976) but in the fall of 1974, when the fish were expected to return, morpholine was not added to Oak Creek. The results of this experiment were different from the others; exposed and non-exposed fish were captured in equally low numbers (51 vs 55) at about the same rate as control fish from previous experiments (Table 6.3). This experiment illustrates the importance of morpholine to the return of imprinted salmon.

In a more refined test, one group of coho smolts held at hatcheries outside of the Lake Michigan drainage basin was exposed to morpholine, a second group to phenethyl alcohol (PEA, $C_8H_{10}O$), and a third left unexposed (Scholz *et al.* 1978b). All three groups were released in Lake Michigan midway between two test streams that were located 9.4 km apart (Figure 6.3). The artificial imprinting for these experiments was done in the spring of 1973 with 5000 fish in each group and repeated in the spring of 1974 with 10 000 fish in each group. During the spawning migration 18 months later, in the fall of 1974 and again in the fall of 1975, morpholine and PEA were metered into separate test streams — the Little Manitowoc and East Twin Rivers, respectively (Figure 6.3). The streams were surveyed for marked fish by creel census, gill-net fishing, and electrofishing. Other locations were also monitored to determine if significant numbers of imprinted fish were straying into non-scented streams. The results from both experiments (Table 6.4) show that the majority of the fish exposed to morpholine were captured in the stream scented with morpholine and most fish exposed to PEA were captured in the stream scented with PEA. By contrast, large numbers of control fish were captured at other locations.

The results of our artificial imprinting experiments provide direct evidence for olfactory imprinting in coho salmon. We have also conducted three experiments with rainbow trout (Scholz *et al.* 1975,

Table 6.2: Census record of coho salmon caught at Oak Creek in fall 1973.

Experiment	Treatment	Stocking location	Number released	Date released	Number recovered	Per cent of fish stocked
1	Morpholine	0.5 km south of Oak Creek	5000	May 1972	437	8.74
	Control	0.5 km south of Oak Creek	5000	May 1972	49	0.95
2	Morpholine	0.5 km south of Oak Creek	5000	May 1972	439	8.78
	Control	0.5 km south of Oak Creek	5000	May 1972	55	1.10
3	Morpholine	13 km north of Oak Creek	8000	May 1972	647	7.89
	Control	13 km north of Oak Creek	10000	May 1972	65	0.65
TOTAL	Morpholine		18000		1515	8.42
	Control		20000		169	0.85

Source: from Cooper *et al.* 1976

Table 6.3: Census record of coho salmon caught at Oak Creek in fall 1974.

Treatment	Stocking location	Number released	Date released	Number recovered	Per cent of fish stocked
Morpholine	0.5 km south of Oak Creek	5000	May 1973	51	1.02
Control	0.5 km south of Oak Creek	5000	May 1973	55	1.10

Note: Morpholine was not present in Oak Creek

Source: from Scholz *et al.* 1975, and Cooper *et al.* 1976

Table 6.4: Total number morpholine-exposed (M), phenethyl alcohol-exposed (PEA), and control (C) salmon captured at individual locations. Data from the Little Manitowoc River (morpholine scented) and Two Rivers area (PEA scented) are printed in italics.

Location	No. Recovered (1974)			No. Recovered (1975)		
	M	PEA	C	M	PEA	C
1 Stony Creek	1		4			12
2 Annapee River		2	7	6	1	37
3 Three Mile Creek	2	1	1			2
4 Kewaunee River	—	—	—	—	—	—
5 Nuclear Power Plants	1		4			2
6 Molash Creek			2			
7 *Two Rivers Breakwater*	*3*	*118*	*15*	*3*	*192*	*12*
8 *East and West Twin Rivers*		*15*	*7*	*3*	*8*	*21*
9 Stocking Site	1		1			1
10 *Little Manitowoc River*	*207*	*6*	*21*	*452*	*14*	*52*
11 Big Manitowoc River	2	3	31		1	26
12 Fisher Creek			3			1

Source: from Scholz *et al.* 1975

Figure 6.3: Research area, Wisconsin shore, Lake Michigan. Numbers in parentheses represent the number of streams in the general area of the monitoring station that were surveyed. Inset shows detail of the release site, the morpholine-scented Little Manitowoc River (M), control (C) and the phenethyl alcohol-scented breakwater area at Two Rivers (PA).

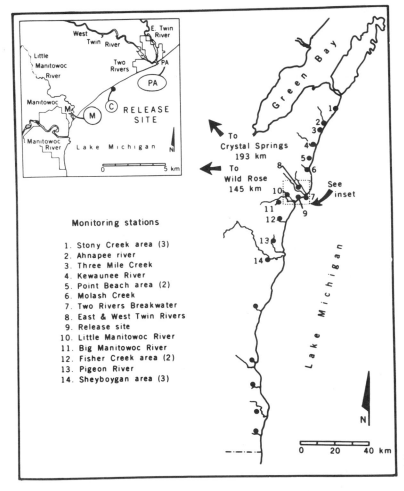

Monitoring stations

1. Stony Creek area (3)
2. Ahnapee river
3. Three Mile Creek
4. Kewaunee River
5. Point Beach area (2)
6. Molash Creek
7. Two Rivers Breakwater
8. East & West Twin Rivers
9. Release site
10. Little Manitowoc River
11. Big Manitowoc River
12. Fisher Creek area (2)
13. Pigeon River
14. Sheyboygan area (3)

1978b; Cooper and Scholz 1976) and one with migratory brown trout (Scholz, Horrall, Cooper and Hasler 1978a) and in all cases significantly higher numbers of morpholine-exposed, as opposed to unexposed trout, returned to a stream scented with morpholine. In a preliminary experiment (Cooper *et al.* 1976), coho salmon that had been exposed to morpholine for two days at the onset of smolting returned to a simulated home stream in about equal numbers as fish exposed for 30 days. Thus, very short periods of morpholine exposure seem sufficient to imprint fish successfully.

Pheromones

Nordeng (1971) has proposed that salmon themselves leave an odour trail (pheromones). He took eggs from one population of sea-char (*Salmo alpinus*) which makes a yearly seaward migration but for only about one month of the summer. The eggs were hatched and reared in a hatchery at Voso, Norway, 1100 km south of Salangen. The sea-char were flown back when they were ready to migrate four years later. One batch was released at the Salangen estuary, and one at another river mouth 10 km west on the same fjord. The majority returned to the Salangen river to which their parents belonged; only a few entered other rivers in the area.

In our Wisconsin rivers, there are no resident salmon to serve as producers of pheromones, hence only the artificial imprinting odour served as the attractive scent. Moreover, the Lake Michigan salmon population is maintained by stocking hatchery raised smolts.

Døving (1979) and Døving, Selset and Thommesen (1980) have presented electrophysiological evidence that char can sense trace concentrations of bile salts (50 µl diluted by 500 000 litres of water) and suggests therefore that the scent of salmon faeces might serve as a cue to returning salmon. Hence, the pheromone hypothesis should be tested further.

Owing to the diversity of odours of rivers to which salmon may become imprinted, the Norwegian suggestion must be acknowledged as a plausible theory. To test this hypothesis, I suggest that a spawning channel be constructed adjacent to a river similar to the spawning enhancement channels built in Western USA. Such a facility in which salmon had not resided would enable a design for rigorous testing in a future experiment.

Summary

Taken together, we believe the results of our recent unified studies provide conclusive evidence for olfactory imprinting. In addition, since our studies were conducted in the field, they provide direct evidence that coho salmon retain and use artificial chemical information to achieve successful homing. It seems likely that salmon possessing this ability would also use it in the natural environment. Pheromones as alternative or additional imprinting scents are discussed.

Our findings have direct practical applications for salvaging endangered stocks of salmon as well as being interesting from a purely scientific viewpoint.

7 HORMONES AND REPRODUCTIVE BEHAVIOUR IN TELEOSTS

N.E. Stacey

Successful reproduction depends not only on the synchronization of gamete production with relevant environmental and social factors so as to optimize the survival of offspring, but also on the synchronization of sexual behaviours with the appropriate stage of gamete development. It has been known for many years that sexual behaviour can be abolished by gonadectomy and restored by crude extracts of the excised organs. The availability of pure steroid hormones and the development of sensitive and specific assay techniques have led to rapid advances in the understanding of how chemical signals from the gonads exert their effects on behaviour. Measurement of blood hormone changes during various reproductive events helps both in assessing the physiological relevance of exogenous hormone treatments and identifying potential hormone-behaviour interactions. Much recent research in mammals and birds has focused on the way in which gonadal hormones released during early development exert powerful influences on future behavioural potential, and how these processes differ among species. Brain lesion and stimulation, brain steroid implants, autoradiography, and electrophysiological and biochemical techniques all have been used to identify specific brain areas mediating the effects of hormones on sexual behaviour. Also it is becoming clear that earlier concepts of reproductive hormones simply exerting relatively long latency effects on behaviour were oversimplified; not only are some of these effects quite rapid, but the interactions of hormones and behaviour are often reciprocal. Today, the study of hormones and reproductive behaviour is a rapidly expanding and multidisciplinary field employing diverse experimental approaches (Hutchison 1978; Beyer 1979). However, this research effort for the most part has focused on mammals; in fish our knowledge of hormonal control of reproductive behaviour is quite limited and based on studies in relatively few species (Liley 1969; Stacey 1981).

The teleost fishes display an impressive diversity of reproductive adaptations (Breder and Rosen 1965; Hoar 1969; Balon 1975). At

117

one extreme are the numerous species which utilize the presumably ancestral vertebrate pattern of oviparity and external fertilization. The simplest pattern may involve essentially promiscuous mating with no premating preparations for egg care or postmating defence of the eggs or young; species reproducing in this way generally produce large numbers of eggs. However, other externally fertilizing fish form pair bonds or establish breeding territories prior to spawning and exhibit elaborate behaviours associated with nest-building and parental care of their relatively small broods. Internal fertilization has arisen many times in the long evolutionary history of teleosts, and today is associated with a range of structural and physiological specializations. In its simplest form, internal fertilization simply ensures the union of egg and sperm prior to oviposition; although sperm may be stored for some time in the female's reproductive tract, oocytes of these oviparous species are fertilized and laid very soon after they are released from the follicles. In ovo-viviparous forms, fertilization and embryogenesis occur either within the follicle or within the lumen of the ovary, and young are released just before or soon after hatching. In this type of reproduction, yolk deposited in the egg prior to fertilization serves as the only source of the embryo's nutritional requirements. On the other hand, the young of truly viviparous species are nourished at least partially by the mother and may be born at a later stage of development, in some cases even when sexually mature.

In addition to this range of adaptations concerned with the timing of fertilization and the handling of fertilized eggs, many teleosts also display less 'conventional' reproductive adaptations. Hermaphroditism, once considered uncommon in teleosts, has now been described in many species (Smith 1975). Variations include simultaneous hermaphrodites, in which both mature sperm and eggs are produced at the same time, and sequential hermaphrodites which may be either protandrous (first male, then female) or protogynous (first female). In some sequential hermaphrodites, transition from one sex to the other can be extremely rapid and may be initiated by social factors such as the absence of individuals of the opposite sex. Although less common than hermaphroditism, parthogenesis has been described in teleosts; here, eggs may begin development spontaneously, or following stimulation by the sperm of a related bisexual species.

Although teleosts thus can be seen to possess an array of reproductive specializations unmatched by any other vertebrate class, surprisingly little research has been concerned with the hormonal and physiological regulation of teleost reproductive behaviour. Obviously, studies of the control of reproductive behaviours in this ecologically and economically important group are of interest and value in their own right; moreover, they can be expected to provide insights into the evolution of hormonal control of sexual behaviour within the vertebrates.

Sexual Behaviour in the Male

Male reproductive behaviours in teleosts encompass a wide range of activities which may include establishment of breeding territories, attraction and courtship of females, release of sperm associated either with copulation or with fertilization of oviposited eggs, and various behaviours involved with care of the eggs or young. However, regardless of the degree of complexity in male reproductive activities, sexual behaviour *per se* — those sexual interactions with the female directly involved with fertilization, whether in externally or internally fertilizing species — always is synchronized with the presence of mature and releasable sperm. It seems likely that this persistent sexual role of the male as 'gamete donor', in fish and in all other vertebrate groups, explains why an evidently primitive hormonal control of male sexual behaviour, increased steroid synthesis associated with testicular maturation, has been retained with little modification throughout vertebrate evolution. In contrast, a major change in the sexual role of the female, from 'donator of oocytes' (external fertilization) to 'recipient of sperm' (internal fertilization), has been accompanied by fundamental changes in the regulation of female sexual behaviour (Stacey 1981).

Steroids and Male Sexual Behaviour

In a variety of teleost species, male sexual behaviour is reduced or abolished either by castration or administration of antigonadotrophins and antisteroids and restored in castrated or drug-treated males by androgen replacement therapy (Liley 1969, 1980). These studies thus demonstrate that testicular androgen plays an important role in the control of male sexual behaviour. In other studies, however, inconsistent effects of these various treatments have led to considerable speculation as to the nature of the hormonal regulation of sexual behaviour in male teleosts.

Some of the confusion evidently is the result of classifying as 'reproductive' those behaviours such as aggression which also may be performed in a non-reproductive context; as emphasized by Liley (1980), agonistic behaviours that are similar in form may well have different causal mechanisms depending on the context in which they occur. In other cases, the purported effectiveness of gonadectomy or drug treatment appears to be unsubstantiated.

For example, when castration has failed to eliminate male sexual behaviour, it sometimes has been suggested that gonadotropin, rather than androgen, may directly stimulate behaviour (Fiedler 1974; Liley 1969, 1980; Crews and Silver 1980). However, sexual behaviour of 'castrates' in many instances may have been due to steroids released by regenerating testes or testicular remnants. Male sexual behaviour in the blue gourami, *Trichogaster trichopterus*, is a good case in point. Castration of male *trichopterus* does not influence non-reproductive aggressiveness, but does lead to a variable decrease both in nest-building and spawning and in the length of the dorsal fin, and androgen-dependent male secondary sex character (Johns and Liley 1970). Of sixteen castrated males tested in this study, nest-building and spawning occurred normally in five males, while these behaviours in the remaining eleven were readily restored by methyltestosterone replacement therapy. Only one of the five males which spawned prior to steroid treatment had identifiable testicular tissue. However, the dorsal fins of the five responsive castrates were longer than those of the eleven initially unresponsive castrates, indicating the presence of endogenous androgen. Similar results have been reported in another belontiid, *Macropodus opercularis*, in which spawning behaviour of castrated males recovers rapidly, coincident with gonadal regeneration. However, the antigonadotropin, methallibure, blocked both regeneration and recovery of male behaviour in *Macropodus*, while normal spawning activity was induced in methallibure-treated males which also received androgen (Villars and Davis 1977). In summary, there appears to be no clear evidence that hormonal control of male sexual behaviour in teleosts differs substantially from what has been demonstrated in other vertebrate groups, i.e. that the onset and maintenance of male sexual activity occurs in response to increased release of steroids from the maturing testes.

The Spawning Reflex in Cyprinodonts

In the oviparous cyprinodont, *Fundulus heteroclitus*, injection of a variety of neurohypophysial hormones (NHHs) induces a 'spawning reflex' response consisting of body flexures and fin movements apparently similar to those normally shown by the male when chasing the female during gamete release (Wilhelmi, Pickford and Sawyer 1955; Pickford and Strecker 1977). Although similar effects have been reported in another cyprinodont, *Oryzias latipes* (Egami 1959), NHHs have not induced this response in a number of non-cyprinodont species (Liley 1969, 1980). The spawning reflex in *Fundulus* is usually seen within thirty minutes of injection, is similar in males and females, and is unaffected by either gonadectomy or hypophysectomy. As pharmacologically high systemic doses of NHH are required to induce the spawning reflex, Macey, Pickford and Peter (1974) suggested that NHHs might induce this behavioural effect by acting within the brain. Although inhibition of the response by lesion of the nucleus preopticus (Macey *et al.* 1974) provided indirect support for this suggestion, more recent studies (Pickford *et al* 1980), showing that intraventricular injection of NHH is no more effective than systemic injection, would indicate a peripheral site of action.

Although the spawning reflex induced by NHHs is distinct and well documented, the relevance of the response to normal spawning behaviour is not apparent. *Fundulus* injected with NHH may display some behaviours associated with normal spawning (Macey *et al.* 1974). However, the spawning reflex is shown by isolated individuals (the standard testing situation in these studies) and evidently involves no coordination between the sexes when injected males and females are placed together. Without further investigation of the endocrine and behavioural factors involved in spawning of *Fundulus*, the possible role of NHHs in the normal sexual behaviour of this and other cyprinodonts cannot be evaluated.

Sexual Behaviour in the Female

Sexual behaviour of female teleosts, unlike that of males, can occur at virtually any stage of gonadal development. In species with external fertilization, female sexual behaviour necessarily occurs after ovulation and is essentially synonymous with oviposition. In internally fertilizing species, female sexual behaviour may occur either after ovulation, as in external fertilizers, or prior to ovulation, apparently the usual condition. Females of some species are able to store viable sperm for considerable periods, and thus may mate at a time when the ovaries are not fully mature. Presumably, it is this temporal lability in the timing of female sexual behaviour with respect to gamete development and fertilization which has led to a greater diversity in mechanisms regulating sexual behaviour in the female than in the male.

The physiological regulation of female reproductive behaviour in teleosts has been examined in only a few species, and any generalizations based on these studies therefore must be viewed with caution. Even so, the present information strongly suggests that the way in which female sexual behaviour is controlled is determined by the timing of sexual behaviour with respect to ovulation (Stacey 1981). Although in some externally fertilizing teleosts (e.g. cichlids) preovulatory female reproductive behaviours occur in those cases where courtship or nest preparation precede ovulation, in the majority of externally fertilizing species the female's repertoire of reproductive activities appears to be limited to postovulatory sexual behaviours associated with oviposition. In internally fertilizing species the situation is not as clear. Preovulatory sexual behaviour obviously must occur in those species in which fertilization is intrafollicular; however, in other internally fertilizing teleosts it is not known whether sexual behaviour precedes or follows ovulation. Much of our present understanding of the regulation of female sexual behaviour in teleosts comes from studies of two species, the internally fertilizing, ovoviviparous guppy, *Poecilia reticulata*, and the externally fertilizing, oviparous goldfish, *Carassius auratus*. Very different mechanisms control female sexual behaviour in these species. However, this is not due to the obvious differences in site of fertilization and post-fertilization fate of the oocytes *per se*, but rather to the fact that sexual behaviour is postovulatory in the goldfish and preovulatory in the guppy.

Postovulatory Sexual Behaviour

Female sexual behaviour in externally fertilizing species can be functional in a reproductive sense only when ovulated oocytes are ready for release. To function effectively, the mechanisms involved must trigger the onset of sexual behaviour soon after ovulation, when the rate of fertilization is maximal, but remain activated only until all ovulated oocytes have been shed. Such a temporal relationship between functional female sexual behaviour and the presence of ovulated eggs is both intuitively obvious and supported by observations in a number of species.

In the goldfish, ovarian growth takes place during the winter and early spring, and ovulation occurs once or several times during the spring breeding season. Ovulation in goldfish is not simply the inevitable consequence of having completed gonadal development, but is a well regulated response to specific environmental cues. Females with mature ovaries will ovulate within several days following an increase in water temperature and exposure to aquatic vegetation (spawning substrate among which the adhesive eggs are scattered). This response involves a dramatic preovulatory increase or surge in blood gonadotropin which begins in the afternoon and induces ovulation during the night (Stacey *et al.* 1979).

Provided that a sexually active male and aquatic vegetation are present, spawning behaviour in the ovulated female commences at dawn on the day of ovulation and continues for several hours until all ovulated eggs are released. As spawning is prematurely terminated by stripping out the ovulated eggs, and restored by simply injecting ovulated eggs through the ovipore and into the ovarian lumen, it is clear that female spawning behaviour in goldfish is synchronized with ovulation by the presence of intraovarian ovulated eggs (Stacey and Liley 1974). Furthermore, the finding that egg injection also reliably induces normal spawning behaviour in females whose ovaries would not have reached an ovulatory condition for several months demonstrates that, although sexual behaviour in female goldfish normally occurs only after ovulation, the female in fact is capable of performing normal levels of sexual activity throughout much of her annual reproductive cycle, provided ovulated eggs are in the ovaries.

The stimulatory action of ovulated eggs on spawning behaviour in goldfish appears to be mediated by a prostaglandin (PG). Both in ovulated females and in non-ovulated females injected with ovulated eggs, spawning behaviour is completely inhibited by the PG synthesis inhibitor, indomethacin (Id), and restored in Id-treated fish by injection of PG (especially PGF_2 alpha) (Stacey 1976, 1981; Stacey and Goetz 1982). The presence of ovulated eggs is not required for PG to stimulate spawning behaviour. In females which have neither ovulated nor been injected with eggs, PG induces female spawning behaviour apparently indistinguishable from that of ovulated fish, except that no oviposition occurs. The simplest interpretation of these results is that oocytes released into the ovarian lumen at ovulation stimulate the synthesis of PG which in turn stimulates spawning behaviour.

Female goldfish with ovaries in any stage of vitellogenesis readily perform spawning behaviour in response to injection of ovulated eggs or PG. However, these spawning responses are modulated in some way by activity of the pituitary-ovarian axis. Injection of ovulated eggs fails to induce spawning either in hypophysectomized females (Stacey 1977), or in intact females with regressed, non-vitellogenic ovaries (Stacey and Liley 1974). In hypophysectomized females, the spawning response to injection of eggs (Stacey 1977) or PG (Stacey 1976) is restored by gonadotrophin treatment. In contrast, steroid treatments have been completely ineffective in restoring responsiveness to eggs or PG in hypophysectomized females. Although oestradiol (Stacey and Liley 1974) and other steroids (Stacey 1977) do restore the spawning response to egg injection in intact, regressed females, these results need not imply a direct action of exogenous steroids in goldfish spawning behaviour. In salmonids, oestrogen and other steroids have been shown to stimulate synthesis of pituitary gonadotrophin (Crim *et al.* 1981); if exogenous steroids have a similar effect in goldfish, they could increase sexual responsiveness indirectly by stimulating endogenous steroid synthesis. Thus, while female spawning behaviour in the goldfish is somehow regulated by gonadotropin and/or steroids, it is clear that for much of the annual reproductive cycle the female normally is in a state of potential sexual responsiveness, but becomes sexually active at ovulation in response to the presence of ovulated eggs.

The effects on female sexual behaviour of removing or injecting ovulated eggs have been studied only in the goldfish. However, from the results of a number of unpublished studies, there is reason to believe that female sexual behaviour in a variety of externally fertilizing species is stimulated by increased PG synthesis following ovulation. For example, as in the goldfish, Id is able to block spawning behaviour in ovulated Pacific herring, *Clupea harengus pallasi* (Stacey, unpubl. obs.), and nest-digging and spawning in ovulated rainbow trout, *Salmo gairdneri* (N.R. Liley, pers. comm.). In the three-spined stickleback, *Gasterosteus aculeatus*, a species with elaborate postovulatory sexual behaviour, spawning of ovulated females is inhibited by Id and restored in some fish by PG injection (T.J. Lam, pers. comm.). PG also has been found to stimulate spawning behaviour in non-ovulated *Puntius gonionotus* (N.R. Liley and E.S.P. Tan, pers. comm.), a broadcast spawning cyprinid, and in non-ovulated *Macropodus opercularis*, a species in which the female releases buoyant eggs beneath the male's bubble-nest (T.A. Villars and D. Chapnick, pers. comm.). Also in a cichlid, the brown acara *Aequidens portalegrensis*, PG injection induces oviposition behaviour in non-ovulated females (Stacey, unpubl. obs.). PG-induced oviposition behaviour in *Aequidens* is of particular interest, as female reproductive behaviours in this and other cichlids may commence several days prior to ovulation. There is some evidence that these preovulatory behaviours, which involve courtship interactions with the male and preparation of the spawning site, are stimulated by steroids (Liley 1969). Thus, cichlids, which spawn readily in the laboratory, may prove to be ideal species in which to investigate how the sequential effects of steroids and PG might synchronize female reproductive behaviours with ovarian maturation and ovulation.

PG also has been found to stimulate female sexual behaviours in two externally fertilizing anuran species. In the leopard frog, *Rana pipiens*, non-ovulated females emit a release croak which inhibits clasping attempts by the male. Release croaking is inhibited in non-ovulated females in which the abdomen is distended by water accumulation induced either by cloacal ligature or injection of arginine vasotocin (AVT) (Diakow and Raimondi 1981). Id reduces the effect of AVT on release croak inhibition, while in non-ovulated females PG injection rapidly inhibits release croaking (Diakow and Nemiroff 1981). PG injection also induces receptive behaviour in non-ovulated *Xenopus laevis* (D.B. Kelley, pers. comm.). These findings indicate that, in a wide variety of externally fertilizing female vertebrates, PG may play an important role in stimulating female sexual behaviours which occur after ovulation.

Although it is clear that exogenous PG is very effective in stimulating postovulatory female sexual behaviour, it must be emphasized that many aspects of the apparent role of PG in spawning behaviour remain unresolved. At least in the goldfish, the present information is consistent with the hypothesis that PG, synthesized within the reproductive tract in response to intraovarian ovulated eggs, is released into the bloodstream and then acts within the brain to trigger spawning behaviour. Bouffard (1979) has found that in goldfish induced to ovulate with human chorionic gonadotrophin (hCG), ovarian fluid bathing the ovulated eggs contains high levels of PGF. Significantly, the concentration of PGF in the blood of hCG-treated fish remains low until ovulation, at which time blood levels increase dramatically and spawning behaviour commences. By twelve hours after ovulation, blood PGF levels remained high in females which retained eggs in the ovaries, but decreased to preovulatory levels in those fish from which the eggs had been stripped at the time of ovulation. These findings thus not only show that high blood levels of

PGF occur at the time when female goldfish are sexually active, but also indicate that the continued presence of eggs in the ovaries maintains PG synthesis.

That PG does not stimulate spawning behaviour through some action on the reproductive tract is suggested by the finding that removing the posterior portions of the ovaries, the oviduct, and the ovipore and surrounding tissues does not alter the female's responsiveness to PG injection (Stacey and Peter 1979). Furthermore, as PG is more effective in inducing spawning when injected into the third ventricle of the brain than when injected systemically, it appears that PG exerts this behavioural action within the brain. Together, these studies in goldfish would suggest that, in inducing spawning behaviour, PG acts as a hormone, a blood-borne chemical messenger triggering a distinct response in a distant target organ. This interpretation must be viewed with caution, however, as in almost all their other functions PGs are known to act as local intracellular or extracellular regulators, and not as hormones in the classical sense. Indeed, were it not for the evidence that blood PG levels are elevated at ovulation, a more acceptable interpretation of the present information might be that spawning is triggered by an increase in brain PG synthesis in response to afferent neural input from the ovulated ovary.

Regardless of where PG involved in spawning behaviour might be synthesized, the way in which PGs stimulate spawning behaviour indicates these substances are well suited to synchronizing female sexual activity with the presence of ovulated eggs. Female goldfish can begin to perform spawning behaviour within several minutes of PG injection and similar rapid responses have been observed in *Macropodus* and *Puntius*. These short latency actions of PG are consistent with observations that female spawning behaviour in a number of teleosts begins very soon after ovulation. At least in the goldfish, the behavioural response to PG is quite sensitive, some females responding to intramuscular doses of PGF_2 alpha as low as 1 $ng\,g^{-1}$ body weight. At higher doses, both the frequency and duration of spawning are increased, although even at doses which induce spawning frequencies typical of ovulated females, the response generally is terminated within several hours. Metabolism of the injected PG, rather than some change in the female's responsiveness following injection, appears to be responsible for the brief duration of PG-induced spawning; females perform progressively less spawning as the interval between time of injection and testing is increased, whereas two injections given several hours apart elicit equivalent responses (Stacey 1981).

An intriguing aspect of PG-induced female spawning behaviour in the goldfish is that PG injection in males induces female behaviour which apparently is indistinguishable from that shown by the female. PG-treated males are not inhibited from performing male behaviour and, when placed with both male and receptive female partners, will rapidly alternate between performance of homotypical and hetero-typical sexual behaviours. This responsiveness to PG in male goldfish provides valuable opportunities to examine, within the individual, both the neural substrates and hormonal stimuli regulating male and female behaviours. In addition, this finding suggests a likely physio-logical basis for the behavioural bisexuality of simultaneous hermaphrodites.

Preovulatory Sexual Behaviour

The physiological control of preovulatory sexual behaviour in teleosts apparently has been studied only in the guppy, a species in which fertilization is intrafollicular. In an extensive series of experi-ments, Liley and his colleagues have clearly demonstrated not only that the endocrine system regulates female sexual behaviour in the guppy, but also that courtship experience plays an important role in modulating the level of female responsiveness.

Sexual behaviour in the female guppy is cyclical and well syn-chronized with a cycle of ovarian development. Sexual receptivity (the tendency of the female to respond to male mating attempts and engage in copulation) is maximal for several days after parturition, coincident with peak ovarian steroidogenic and vitellogenic activity (Lambert and Van Oordt 1974). Following copulation and im-pregnation, the female becomes unreceptive to the male and remains so until parturition, at which time receptivity again increases.

Ovariectomy abolishes receptivity in sexually experienced female guppies, while oestradiol and other oestrogens effectively restore sexual responsiveness following gonadectomy (Liley 1972). Gonadotrophin injections also restore sexual responsiveness of hypophysectomized females. However, as gonadotrophin is beha-viourally effective only if the ovaries are present (Liley and Donaldson 1969), whereas oestrogen restores receptivity even in females which are both ovariectomized and hypophysectomized (Liley 1972), it appears that gonadotrophin influences sexual behaviour in the guppy only indirectly, by stimulating ovarian oestrogen secretion.

The early post-partum period in the guppy not only is the time of maximal sexual responsiveness but also the time when the female is most attractive to the male; water from early post-partum females is more attractive to males than is water from females in the middle of the gestation cycle (Crow and Liley 1979). Furthermore, courtship in male guppies is stimulated by water from hypophysectomized females treated with either gonadotrophin or oestrogen, but not by water from ovariectomized, oestrogen-treated females (Meyer and Liley forthcoming), suggesting that oestrogen, acting on the ovary to induce pheromone release, serves to synchronize attractivity with receptivity at the time of follicular maturation. There also is much evidence that, in a variety of oviparous teleosts, substances in the ovarian fluid of ovulated females may perform a pheromonal function in stimulating male courtship (Liley forthcoming); however, the possible involvement of hormones in this pheromone production apparently has not been investigated.

In contrast to sexually experienced female guppies, in which sexual receptivity is well synchronized with the ovarian cycle, naive virgin females tend to be highly receptive when first exposed to male courtship (Liley and Wishlow 1974). Repeated exposure to males, even to males which are gonopodectomized and therefore incapable of insemination, leads to a rapid reduction in receptivity. Neither the high initial levels of receptivity in naive virgins, nor the decline in receptivity induced by courtship, requires the presence of the ovaries. Even if they had been ovariectomized as much as twenty-four days prior to testing, naive virgins displayed typically high levels of receptivity during initial courtship encounters. However, whereas intact virgins, when tested repeatedly with gonopodectomized males for up to six weeks, showed transient recovery of receptivity suggestive of an endogenous cycle of ovarian activity, ovariectomized virgins remained unreceptive on repeated testing. Liley and Wishlow suggest that the initially high levels of receptivity in virgin females may ensure insemination on the first exposure to a male regardless of the stage of ovarian development, while rapid habituation of this responsiveness following coitus may serve to reduce the impregnated female's exposure to predation. This demonstration that sexual responsiveness in the female guppy is rapidly attenuated by courtship experience, is consistent with studies of a variety of internally fertilizing female vertebrates (references cited in Tokartz and Crews 1981) in which the sequential stimulatory effects of ovarian steroids and the inhibitory effects of coitus serve to restrict sexual receptivity to the periovulatory period.

Summary

Certainly considerable progress has been made in understanding the physiological mechanisms which synchronize sexual behaviour with gonadal maturation in fishes. Sex steroids evidently play important roles in regulating female sexual behaviour in the guppy, and male sexual behaviour in a number of species. Specific brain areas which concentrate sex steroids have been identified in several teleosts, although stimulation of spawning or other reproductive behaviours in teleosts by a direct action of steroids on the brain has yet to be demonstrated (Demski and Hornby 1982). Prostaglandins have been shown to exert potent and rapid stimulatory effects on female sexual behaviours in a variety of oviparous teleosts; however, it is not clear whether these behavioural actions of prostaglandins are indicative of a truly hormonal role of prostaglandins released into the circulation, or whether female spawning behaviour may simply be triggered by prostaglandin perhaps synthesized within, and acting on, the brain. Thus, while foundations have been laid in several areas, much more information is required to provide a comprehensive understanding of the physiological regulation of sexual behaviour in even those few species which have received most attention.

Despite the many important questions still to be resolved, studies of reproductive behaviour in fish already have provided the basis for a broader perspective of hormones and reproductive behaviour. The steroid dependence of male sexual behaviour, and of female behaviour in internally fertilizing species, is a phenomenon widespread among the vertebrates. Similarly, the mechanism regulating female sexual behaviour in externally fertilizing teleosts appears not to be unique to this group, but also may occur in those anurans which face similar problems in synchronizing sexual behaviour with the presence of ovulated eggs.

8 INTEGRATION AND CONTROL BY THE CENTRAL NERVOUS SYSTEM

D.M. Guthrie

The central nervous system controls the behaviour of an animal by directing its movements in accordance with the prevailing pattern of sensory information and its own motivated state at the time. The endocrine system not only has a profound effect on the way in which the motivational functions occur, but also regulates the economy of the body in an appropriate manner.

Professor Dodd (to whom this book is dedicated) has unravelled for us many of the secrets of endocrine systems in the lower vertebrates, and the emphasis in this volume is quite rightly on this aspect of control physiology. In this chapter I shall attempt to concentrate more on the sensory aspect of nervous function, and on the transmission of information by short-term electrically-mediated processes rather than by hormonal means.

If we look at a transverse section through the midbrain of a teleost fish we are struck by the immense complexity of its structure. Even a small area of the section, if suitably stained, reveals hundreds of nerve fibres arranged in complex patterns, each physically distinct pathway capable of making a unique contribution to ongoing control functions. An electronmicrograph of one of the more densely structured areas might show as many as 50 minute fibre profiles within a single micron square. This tells us that an extraordinarily complex structure is needed to maintain and direct behaviour under the apparently rather undemanding conditions for life in the stable environments inhabited by many fishes. Modern electrophysiological techniques enable us to follow the individual activity of a cell or a fibre, as well as to monitor the behaviour of neuron groups. Some examples of these and other methods will be described in this chapter.

Control Processes and the Fish Central Nervous System

Despite their comparatively simple structure, teleost fishes rival mammals and birds in their evolutionary success in terms not only of numbers of species, but in the range of their behaviour. Indeed, some degree of convergent evolution is apparent in the elaborate courtship displays exhibited by both birds and teleost fish. Yet the central nervous system, and in particular the brain of fishes, is relatively much smaller than is it in higher vertebrates. Jerison (1973) quotes a figure of 0.005 per cent body weight for eels and coelocanths, and this can be compared with a range of 0.5—2.0 per cent for most birds and mammals. We have very little indication of comparative cell numbers, but estimates of the cell complement of the optic tecta of the goldfish by Meek and Schellart (1978) suggest values of 4—5 million cells. The optic tectum is the primary visual projection area. For comparison the primary visual area of the cerebral cortex (area 17) in the macaque would contain approximately 67 million cells (Blinkov and Glezer 1968).

To understand how such an economical system can provide a range of complex and adaptively appropriate functions we must briefly discuss the general nature of these functions.

It has been usual to describe the functions of the nervous system by the terms integration and control. These words are not inappropriate but they do imply an approach that has changed with better understanding of nervous function. Integration suggests a greater degree of convergence within the sensory system as a whole (to produce a particular motor outcome) than I believe is of general occurrence. It involves the idea of the nervous system as a reflex machine, and plays down the importance of information stored in the nervous system. Control could be taken to refer simply to the peripheral motor system. Both integration and control of movement are simply elements in the broader strategies of the nervous system. These strategies can be enclosed within the general concept of homeostasis, and can be compared with the more obvious examples of homeostatic reflexes. The appetitive drive triggered by hunger that causes an animal to forage actively and selectively to restore blood metabolite levels, can be compared with the vasodilator reflex controlling body temperature, both involving centres in the hypothalamus. The difference lies in the amount and quality of the information required, the complexity of the motor sequences involved, and the variability of the pathway to equilibrium. In fact the overall foraging strategy may be much the same on different sorties, but it consists of many small steps that can be varied according to the probability of their success. At this point we can begin to contrast the simple examples of homeostasis like the vasomotor reflex, where the response is directly proportional to input, the response type is invariant, and the probability of a successful outcome near 100 per cent; to the foraging sortie triggered rather than progressively controlled by the input signal, varied according to circumstances, and with a probability of success that for a predator may be less than 10 per cent. This also suggests that a predictive capability based on memory is usefully involved in the more adventurous strategies, and that this entails the collection of much more information than will be used on any particular occasion. We have no evidence that this process by itself requires a particularly large number of nerve cells, but two processes that involve the quality of the information probably do. These are both necessitated in signal recognition, and they are (a) contrast enhancement for the reduction of ambiguity and (b) convergent processing, again connected with stimulus ambiguity.

The Evolution of the Brain in Fish

The subdivision of the brain according to the major groups of sensory fibres — (dorsal lobes) and the endocrine (diencephalon) and motor centres (medulla), can be traced in many fish (Figure 8.1).

The direct way in which an increased development of a sensory component results in the development of dorsal lobes can be clearly seen in cyprinids in the formation of vagus lobes (only found in some forms) in relation to gustatory function.

Despite the importance of auditory functions in some fish, the processing of water-borne sound stimuli seems to have led not so much to the development of external dorsal lobes, but rather to deep lying nuclei, and dorsal auditory lobes are rarely visible, although they can be seen in the tunny (Figure 8.2).

The trend in modern teleost fishes seems to have been towards the development of a massive central region of the brain composed of the tectum and tegmentum of the midbrain underlain by diencephalic regions, complex thalamic and hypothalamic nuclei and tracts, and penetrated by a large extension of the cerebellum, the valvula cerebelli. These structures are shown diagrammatically in Figure 8.1. This central brain region can be compared to the comparatively small, and in part histologically rather simple, forebrain. Aronson (1981) has sounded a useful note of caution over regarding the forebrain as a simple olfactory centre, and there can be little doubt now that important modulation of motivational states depends on cell

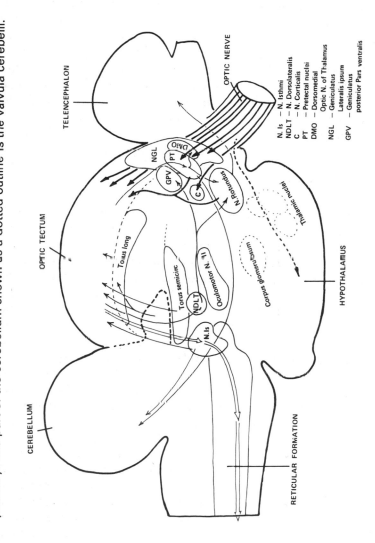

Figure 8.1: Diagram of the teleostean brain as seen from the side, with particular emphasis on the visual pathway. The part of the cerebellum shown as a dotted outline is the valvula cerebelli.

TELENCEPHALON

OPTIC NERVE

OPTIC TECTUM

CEREBELLUM

RETICULAR FORMATION

HYPOTHALAMUS

NGL

DMO

PT

GPV

C

N. Rotundus

Torus long

Torus semicirc

NDLT

Oculomotor N. II

N. Is

Corpus glomerulosum

Thalamic nuclei

N. Is — N. Isthmi
NDLT — N. Dorsolateralis
C — N. Corticalis
PT — Pretectal nuclei
DMO — Dorsomedial
         Optic N. of Thalamus
NGL — Geniculatus
         Lateralis ipsum
GPV — Geniculatus
         posterior Pars ventralis

groups in the forebrain. Segaar's classical study on the stickleback illustrates the existence of forebrain cells influencing in a positive or negative manner the likelihood of courtship or conflict behaviour (Segaar 1961).

In the same way we recognise that only some of the outer layers of the optic tectum are purely visual and that the lower layers receive fibres from many other brain regions and are not diminished by ablation of the eye, although they do contain visually excited cells (Guthrie and Banks 1978). The stratum marginale has been shown by Ito (1970) and Vanegas, Williams and Freeman (1979) to contain fibres with cerebellar connections that project to the tectal pyramidal cells. These fibres are believed to originate in the torus longitudinalis and are not affected by eye removal.

It is clear from this that the dorsal lobes of the teleost brain are not exclusively devoted to particular sensory functions, but certain pathways that are functionally related in a rather broad way, and perhaps are associated for both motor and sensory reasons.

Some light on the relation between brain development and functional specialisation may come from comparison of the brains of different fish species. Tuge and Uchihashi (1968) in their monumental work on the brains of fishes provide us with some information concerning the interspecific differences to be observed in the brain structure of fishes with different habits. Some trends do appear. Large-eyed plankton feeders or fish predators do have large optic lobes. Examples are *Konosirus* and the tunny *Thynnus*, two marine species with folded tectal lobes. In forms that hunt largely by smell, llike the black porgy (*Acanthopagrus*) the olfactory bulb and the telencephalic lobes are relatively large, and the optic tectum is smaller. Most fish have a large corpus cerebelli, and this has always been associated with the need for a precise control of balance and posture when swimming in water, since the needs of orientation and locomotion may conflict. Maintenance of orientation in environments with complex and powerful currents is also a special requirement in many species. In the tunny the cerebellum is a large subdivided structure, with a corpus cerebelli extending forward to the telencephalon, and this is presumably associated with an ability to swim at very high speeds. The most extreme development of a brain lobe is found in the mormyrid electric fish, in which again it is the corpus cerebelli that is enlarged (Figure 8.2 illustrates a side view of the brain, and some of the other fish brains mentioned above). Here the great development of the cerebellum is probably associated with

Figure 8.2: Sideviews of the brains of four teleost species to show the varying development of the major lobes. (A) *Acantrhopagrus*, an inshore nocturnal hunter. (B) *Konosirus*, a pelagic plankton-feeder relying on vision. (C) *Thynnus*, an oceanic pursuit-predator of other fishes. (D) *Gnathonemus*, a mormyrid relying on an electroreception system for obstacle avoidance in turbid water. In A the forebrain lobes are enlarged in relation to olfaction, whilst the optic tectum is the major lobe in B. Both the cerebellum and the optic tectum are very well developed in *Thynnus*. The great enlargement of the valvula cerebelli in D is believed to be in connection with the analysis of information from the electroreceptors. acl — acoustic lobe; cbl — cerebellum; ob — olfactory bulb; ot — optic tectum; t — telencephalic lobe; v cbl — valvula cerebelli; v l — vagus lobe.

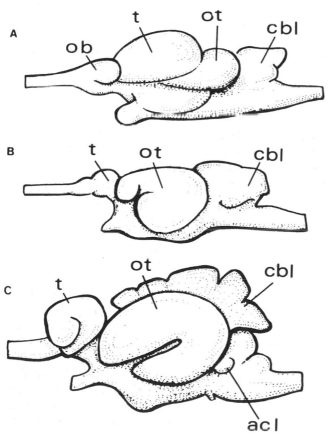

D

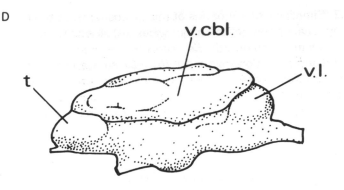

Source: A—C after Tuge and Uchihashi 1968

the analysis of complex disturbances of the pulse field created around the fish by the electric organ (Bullock, Hamstra and Scheich 1972).

How do new developments of this kind arise? It had been thought by students of neural evolution that radically new structures could arise in the brain, and by expansion exclude older regions. Ebbesson working to a great extent with fish brains has concluded that differentiation and development of cell groups and a redistribution of their inputs can account for the changes observed (Ebbesson 1980).

Neuronal Morphology and Tissue Organisation

Neurone Types and Brain Histology.   It might well be thought that the form of central neurons varied relatively little between the different classes of vertebrates. Indeed, some cell types, for example the Purkinje cell of the cerebellum, can be readily identified in both fish and mammal by its candelabra-like dendritic tree. At the same time, I have listed below a number of fairly consistent differences that must be related to functional emphases.

(1)   Many brain neurons in fish possess a relatively isolated cell body, offset from axon and dendrites rather than placed at the geometrical centre of a radiating system of dendrites and axon as in most mammalian neurons. Class 14 tectal cells (Meek and Schellart 1978) and many of the cells in the tegmentum (torus semicircularis) are examples of this physical type (Figure 8.3), which resembles some arthropod neurons. The significance of this feature is uncertain, but it has often been thought likely that a central cell body allowed the major synthetic centres of the cell to be exposed to the varying

Figure 8.3: Different morphological types of neuron from the fish central nervous system. (A) an intraneuron from the torus semicircularis of the trout. (B) Deep horizontal neuron from the tectum of the perch (courtesy of Al Akell). (C) Vertical tectal cell (type 14) from the Jewelfish (*Hemichromis*). (D) Mauthner cell from the Amazon Molly (*Poecilia formosa*). l.d. — lateral dendrite; v.a. ventral dendrite; s — soma. An arrow indicates the process believed to be the axon.

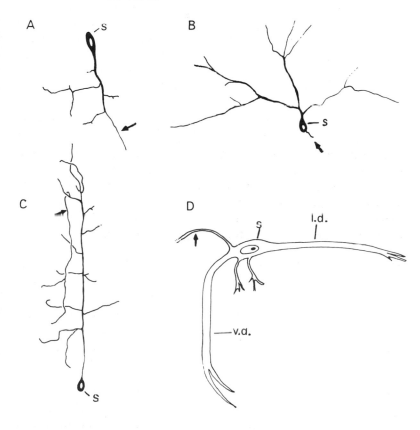

patterns of impulse traffic passing across the neuron, and this might enhance memory functions.

(2)   The neuron-poor nervous systems of arthropods and molluscs are distinguished by the presence of large single identifiable cells. These perform rather specific functions and are identifiable at an early developmental stage in the embryo. They are not a general feature of the higher vertebrate nervous system, but they persist in fish. The Mauthner cell is the most well known, but there are a number of other identifiable single cells forming part of the reticulo-spinal system (including several pairs of Müller cells) in both lampreys and teleosts. Recent studies have thrown some doubt on the uniqueness of function of the giant pair of Mauthner cells (Mauthner cell 1). At one time thought to be the essential link in the auditory-startle response, but Kimmel, Eaton and Powell (1980) have shown that something similar to the startle response survives after Mauthner cell ablation. Nevertheless, the survival of such specialised single cells may be accounted a primitive feature.

(3)   The degree of stratification that occurs anywhere in the fish brain is strictly limited. Although the optic tectum of fishes has a laminated or stratified appearance, this is to a large extent due to layers of fibres, rather than to the precise alignment of neuron cell bodies. In a number of mammals, most notably in primates, as many as seven strata may be visible in part of the cortex, and the hippocampus and some of the deeper nuclei such as the lateral geniculate are often distinctively stratified.

Dependence of Behavioural Acts on Massed Sensory Inputs

Many lines of evidence suggest that fish, in common with other animals can discriminate quite subtle differences in stimuli, and that highly specific stimulus configurations may be needed to trigger particular behavioural acts. Two recent studies, however, demonstrate that under certain conditions, activity in large bundles of afferent fibres seem to be associated with distinctive modes of behaviour, or can alter the level of intensity at which it normally occurs.

In some teleosts, such as cod (*Gadus morhua*) the olfactory tract forms a nerve-like structure running forwards from the brain to the olfactory organ on each side. Each olfactory tract is separated anatomically into four separate strands or nerve fibre bundles containing many thousands of axonal afferents. Døving (1982) found that weak electrical stimulation of one of these bundles would evoke distinct types of behaviour in a freeswimming cod when the other three types of bundle were cut back.

(1)   The stimulation of the outer bundle of the lateral tract produced a head-down swimming position, rather like the one adopted in feeding or foraging behaviour.

(2)   Stimulation of the inner part of the lateral tract produced a high level of swimming activity and a snapping of the jaws that could also be related to feeding behaviour.

(3)   Stimulation of the outer bundle of the medial tract caused the fish to adopt a head-up position, and a quivering of the flanks similar to movements performed during reproductive behaviour.

(4)   The inner part of the medial tract when stimulated seemed to be responsible for the fish pressing itself to the substrate and changing colour to a characteristic pattern of flank stripes, and dorsal spots. Thus quite precise types of behaviour seem to result from the massed activity of a large group of fibres coupled with the removal of other similar inputs.

The results of de Bruin's study on stickleback cranial nerves (de Bruin 1982) is perhaps slightly less striking in that natural stimulation was relied on. This produced a patterned response, and alterations in behavioural intensity rather than presence or absence.

The 9th and 10th cranial nerves are accessible to surgery and have two anastomoses, a dorsal and a ventral one, and fibre bundles separate from the main nerves at these points. Section of the 10th nerve bundle that forms the ventral anastomosis reduces the level of parental (fanning) behaviour. Section of the main root of the 9th nerve prevents the normal sharp suppression of courtship behaviour on egghatching day; whilst removal of the 9th nerve bundle that forms the dorsal anastomosis results in a decrease in courtship behaviour.

The changes in the levels of behaviour are 50-90 per cent different from controls at the maximum point of separation.

Functional Localisation in the Teleost Brain

Attempts to find areas which have a special relationship with particular types of behaviour as evidenced by electrical stimulation and lesioning experiments have increased in number since the early studies of Janzen (1933) and Akert (1949).

As in higher vertebrates some of the most precise effects concern the hypothalamic areas of the midbrain. Stimulation by means of trailing wire electrodes implanted into an area below the nucleus glomerulous in the sunfish *Lepomis* (Figure 8.4) was shown by Demski and Knigge (1971) to produce feeding behaviour. Similar experiments on *Tilapia* (Demski 1973), and on the goldfish by Savage and Roberts (1975) produced comparable results. By recording from single units in the hypothalamic feeding area (HEA) Demski (1981) was able to show that they could be excited by shocks applied to the largely gustatory vagus lobe, or by substances placed in the mouth. Evidence of a satiety centre analogous with the one in mammals has not so far been demonstrated.

Rather more diffuse effects are associated with the forebrain. Forebrain lesions or ablation modify, reduce or disrupt a wide variety of types of behaviour including reproductive behaviour, aggression and learning (Flood and Overmeir 1981). One of the studies that indicated rather localised functions was carried out by Kyle and Peter (1979). They showed that lesions in the ventral and supracommisural parts of the ventral telencephalon (Vv-Vs of Northcutt and Bradford 1978) decreased the spawning ability of male goldfish. The classical study of Segaar on telencephalic influences on breeding behaviour in male sticklebacks revealed a balance between the action of the anteriolateral and the posteroventral parts of the forebrain.

Some of the results of tectal stimulation will be outlined below, but these and other studies indicate that there is indeed about as much functional localisation as one would expect to be able to demonstrate with current stimulation techniques, given the rather small size of the fish brain and the pervasive distribution of some of the neural structures involved. There is at present no reason to believe that functions are any more diffusely organised in fish than they are in mammals.

Methods for Studying the Electrical Activity of the Fish Brain

Most of the electrophysiological techniques currently available have been applied to the study of the fish nervous system. This very brief survey is intended simply to point to the different approaches that have proved successful.

Setting-up Procedures.   One of the problems that we have encountered with perch tectal cells is that the procedures involved in

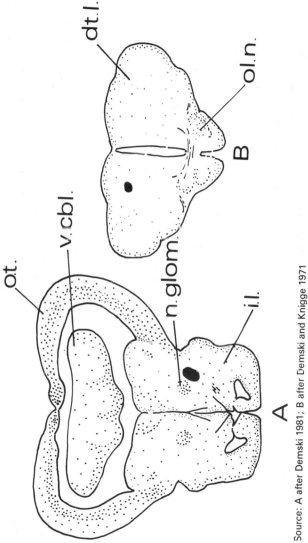

Figure 8.4: Functional centres in the fish brain demonstrated by focal electrical stimulation. (A) Transverse section of the midbrain of the goldfish drawn schematically to show the area (in black) associated with feeding. (B) Transverse section of the telencephalon of the bluegil l to show the site at which electrical stimulation evokes nestbuilding. Dt.l — dorsal telencephalic lobe; i.l., — inferior lobe of the hypothalamus; n.glom. — nucleus glomerulosus; ol.n. — olfactory nerve; o.t. — optic tectum; v.cbl. — valvula cerebelli.

Source: A after Demski 1981; B after Demski and Knigge 1971

catching and handling the fish prior to anaesthesia and surgery produce a strong stress effect and this seems to result in a rather depressed state of excitability of the brain neurons. The period of most intense depression coincides approximately with the 4 hour duration of major effects on blood constituents observed by Wedemeyer (1972) in stressed trout and salmon, but he found that detectable symptoms may last for many days.

It is not feasible to record the responses of central neurons in deeply anaesthetised fish, and surgical isolation of the midbrain or hindbrain by section of the medulla, or forebrain ablation are necessary in the UK to comply with Home Office regulations. We have assumed that this does not greatly affect the properties of many of the central neurons, but this is not known.

Fish can be maintained in air for long periods provided that the gills can be kept constantly perfused with aerated water and the surface of the body is kept moist. Although we usually monitor heart beat rate and regularity with ECG electrodes, this is a fairly insensitive indication of the condition of the nervous system. Microscopic scrutiny of the blood flow in superficial cerebral blood vessels helps in assessing the state of the brain.

The Recording of Massed Potential Waves (EEG, Tectal Evoked Responses) from the Brain.   Using high grain amplifying systems it is possible to obtain well-differentiated records of fluctuations in voltage from the fish brain with simple wire electrodes. Frequencies range from 4—16 Hz in the resting state, to 30 Hz in the stimulated state (Laming 1981). The origin of these slow waves is still rather obscure, but is likely to derive from non-spiking cells, and it is suggested by Elul (1972) from post-synaptic potentials (see Figures 8.5 and 8.6).

Extracellular Unit Recording.   'Proximity records from single neurons still provide the most useful information for the interpretation of behavioural events. Insulated metal microelectrodes and electrolyte filled glass pipettes have both been used successfully for this purpose. Recording from single axons or axon terminals, as for example in the optic nerve and its tectal radiation seems to have been most successful using either insulated metal microelectrodes (Wartzok and Marks 1973) or relatively coarse glass pipettes (4 MΩ resistances with 2.5 M NaCl; Schellart and Spekreijse 1976). However, we have found that only relatively fine glass micro-

Figure 8.5: Electrical responses from fish neurons. (A)
Intracellularly recorded action potential from a molly Mauthner
cell in response to an electrical shock (note shock artefact
preceding a.p.). (B) Spontaneous potential waves recorded
extracellularly from a unit in the binocular quadrant of the goldfish
tectum. Note very long duration. (C) Spontaneous bursts of action
potentials recorded intracellularly from a tectal neuron in the
perch displayed on two time scales. At the start the trace is on the
zero potential level and jumps to the resting potential level when
the electrode penetrates the cell. (D) Pseudoamacrine cell from
the perch tectum responding to light flashes in front of the eye.
intracellular recording.

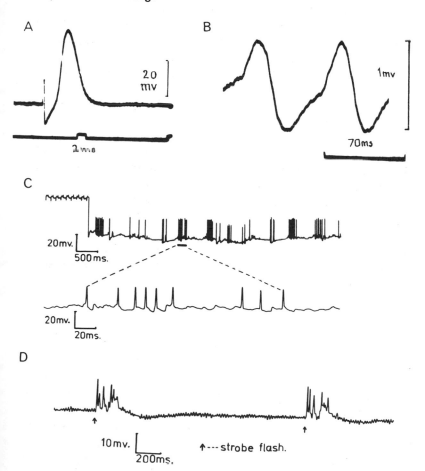

Figure 8.6: The function of retinal efferents. Fibre bundles in the optic nerve respond to a light flash with a rather complex double burst as seen in the top trace (A). After tectal ablation (B) the second burst starts earlier and the preceding slow wave is larger. This was taken as evidence of an inhibitory component in the function of tectal efferents. The traces shown are computer averaged from 64 responses in each case.

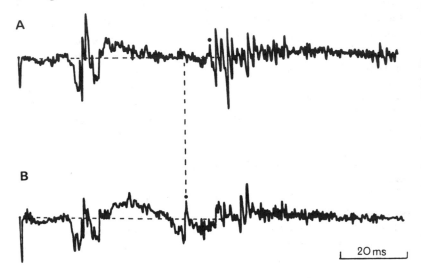

electrodes gave good recording contacts with the heavily myelinated axons in the optic nerve of the perch.

For recordings from intrinsic brain cells, glass pipettes with resistances in the 10-30 MΩ range (3 M KCl) have been found most satisfactory in terms of signal-to-noise ratio (Sutterlin and Prosser 1970; Vanegas *et al.* 1974) although metal electrodes have also been used (O'Benar 1976).

Intracellular Recordings.   For the study of synaptic events, and the unequivocal identification of morphological types of neuron by dye injection, intracellular penetrations are required. By using automated stimulus and data management, and some of the more recent dyes, like Lucifer Yellow, cells need only be held in stable contact for 10-30 seconds for useful information to be obtained. This is important where small cells in the dorsal lobes of the brain are concerned, as these areas are liable to very slow subsidence. For the larger neurons in the reticulospinal formation, and diencephalon intracellular penetrations of neuron somata are relatively easy to achieve, and quite coarse pipettes with resistances of 20-30 MΩ are successful (see Figure 8.5).

Problems arise when attempts are made to introduce electrodes into cells less than 15 μm in diameter. Much finer electrodes are generally advisable (30-100 MΩ, 3M KCl equivalent), but part of the problem lies in the passive mechanical properties of these small neurons and their surrounding membranes which prevent penetration.

One of the most successful attempts at injecting dye into small fish neurons and relating neuron type to functional response has recently been published by Niida and his colleagues (Niida, Oka and Iwata 1980), and the reader should consult his paper for technical details.

Types of Signal Transmission in the Fish Brain

If the faster actions of the fish nervous system depend on transmission of electrical signals from neuron to neuron it may be asked whether these are in any way characteristic of the fish nervous system.

The most investigated neuron in fishes is the giant Mauthner neuron of the medulla, the largest of a group of reticulospinal cells that control rapid evasion movements through their connections with spinal motor neurons. These connections are descending axons along which propagated action potentials pass at up to 100 metres per second. The soma and dendrites of the M-cell are somewhat unusual amongst vertebrate neurons in that they are not electrically responsive, and spikes backfired into the cell by antidromic stimulation of the axon decrement away rather quickly in the cell body. Intracellular penetrations of the M-cell seldom reveal spike potentials of more than 50 mV, with a resting potential value of 70-75 mV (Korn and Faber 1978). Rather low values of the amplitude of the action potential are also characteristic of tectal neurons where they range from 10-25 mV in cells that can be effectively hyperpolarised and depolarised. Another feature of the M-cell function is that repetitive spiking can be inhibited by the action of neurons that come close to, but do not make contact with the M-cell, forming a dense mass called the axon cap. These neurons produce a field potential recorded extracellularly as a positive going wave, the external hyperpolarising potential or EHP. In addition the M-cell spike may cause passive hyperpolarisation of other neighbouring neurons. An M-cell response is illustrated in Figure 8.5a.

Bennett and his colleagues (Bennett 1974) have shown that where there is a need for motor neurons to produce synchronous impulses — as in the electric organ of electric fishes or for controlled eye movement, neurons are electrically coupled, rather than by conventional chemical synapses. Electrical synapses are also present between the giant auditory afferents and the Mauthner cell to reduce synaptic delays to a minimum for the startle response.

With suitable electrodes extracellular spikes can be recorded some distance from their neuron of origin, indeed the extracellular spike of the M-cell can be recorded some hundreds of microns away from the cell.

Where extracellular spikes are being recorded from tectal neurons they may appear as a fast negative phase, followed by a shallow positive wave in the upper part of the tectum. Deeper down the same neuron will provide a weak positive wave followed by quite a large slow negative phase. This reversal, which can also be seen in records of massed potentials, was explained by Vanegas (1975) on sink-source theory. Inflow of positive ions near the spike initiating axon (sink) caused negativity, but ions were drawn from the deeper parts of neuron acting as a source, and this region appeared positive at first.

In practice the fast negative phase type of record can often be obtained from the same unit over traverses of several hundred microns; this is a little surprising, as under orthodromic conditions the spike would not have been expected to invade the dendritic tree very far as the diameter greatly exceeds that of the axon. If, however, the spike-initiating zone is a long segment of the dendritic axis abutting on the axonal base, this might allow more effective communication along very long dendrites, and this would tally better with our observations.

The propagated action potentials and the postsynaptic graded potentials (psp) that can often been seen to precede them are the most easily recorded of the electrical phenomena, registered by an electrode advanced into the fish brain, but other types of event also occur. Penetration of single elements occasionally reveals slow waves, 10-20 ms in duration, which occur either spontaneoulsy or as a response to stimuli. These may be of very small amplitude, 1-2 mV recorded intracellularly. Extracellular recordings of units in the goldfish (Guthrie and Banks 1974) located in the anterior area of the tectum revealed units with 50-60 ms duration slow waves (Figure 8.5b). Here a psp + spike configuration may obtain, although this is largely a matter of interpretation.

One of the most striking types of response has been recorded from the perch, and consists of a dual phase slow wave from which spike-like excursions arise (Figure 8.5d). This type of response, which he refers to as type 2D, has also been recorded in nearly identical form by Witpaard (1976) from deep tectal cells in the frog, and is quite similar to the response of amacrine cells in the mudpuppy *Necturus* (Werblin and Dowling 1969). The response of the slow wave cells in the goldfish also finds a parallel with the response of type 3SA tectal cells observed by Witpaard in the frog. A number of explanations of these findings can be made. They are as follows:

(1) They could be recordings from dendritic extensions sufficiently remote from the spike initiating zone for only graded psps to be observed.

(2) The cells may have such low excitability that spike generation has failed.

(3) The recordings may be from glial or other accessory cells that do not spike, and simply follow the action of local neurons.

(4) Alternatively, as Witpaard suggests, there are a large number of short-range non-spiking neurons in the tectum of lower vertebrates, as there are in the retina.

The Teleost Visual Pathway as an Example of Neural Function in Fish

In the great majority of teleosts the eyes are large and well developed, and a transverse section of the head passing through them reveals that they occupy up to half the area of the section at this point. Casual observation of the exploratory behaviour of fish in natural habitats and in aquaria suggest that despite water turbidity, many species depend to a great extent on vision for information about their environment, in much the same way as do birds and mammals. Clear-water species often possess many distinctive patterns of pigmentation, which appear to have the function of species signals, and this is confirmed by work with models (Hay 1978; Stacey and Chiszar 1978; Katzir 1981). Both colour and form seem to be important and while sensitivity to line separation, wavelength interval, and contour angle may be slightly less than in more visually advanced mammals

(Northmore, Volkmann and Yager 1978) similar processes of feature extraction and synthesis are likely to exist.

The Teleostean Eye

The highly refractive spherical lens can be moved eccentrically within the eye cavity rather than along a fixed axis, as in the case of the mammalian lens with its variable geometry. This is probably one of the reasons for the general absence of a pit-like fovea in teleosts. A few forms like the kelpbass (*Paralabrax*) do have such a fovea, but an area fovealis with densely-packed ganglion cells, situated at the temporal binocular focus is more generally found.

In the more recently evolved acanthopterygian teleosts the accommodatory muscles form a complex array of four or more muscles (see Guthrie 1981). These are responsible for large excursions of the lens in transverse and longitudinal planes (Sivak 1973), and allow objects to be brought into sharp focus on any part of the retina. This contrasts with the situation in fish like trout, pike and goldfish where only a single accomodatory muscle is present.

The retina of teleosts has the general form characteristic of vertebrates, but there are one or two important features. Due to the peripherally situated lens, which allows a wide arc of monocular vision, the iris can seldom be 'stopped down' very far. In consequence adaptation to differing light intensities is performed in the retina by movement of the absorptive outer segments of receptors across the pigment layer — the retinomotor reflex.

In many of the predatory species the cones form a very regular mosaic pattern reminiscent of the arthropod compound eye, a feature possibly related to motion detection. The larger cones often appear rather massive compared with mammalian cones, and this almost certainly results from the development of a chromatic discrimination system designed to work at low light intensities (see Burkhardt 1966). A majority of teleostean species examined provide evidence of chromatic discrimination (Northmore *et al.* 1978), with two or three distinct kinds of cone. There are also several morphological types of horizontal cell and ganglion cell. In terms of structural complexity the teleost retina, as evidenced by the detailed comparative studies of Cajal (see Rodieck (1973) for an illustrated translation), seems to be very similar to that of mammals and some features are found in these two groups and not in others.

Retinal Ganglion Cells

Most of the single cell recording studies carried out on the fish retina have been concentrated on the massive cones, or on the equally large horizontal cells, mainly from the point of view of colour coding; there is relatively little information about other cells or other processes, with the exception of the ganglion cells. Recordings can be made relatively easily from ganglion cell axons either at the point where they form the optic nerve, or where they spread out in the optic tectum.

Recordings of the massed potential wave from the optic nerve, following the application of single shocks, indicate the presence of three or four groups of fibres with different conduction velocities (Vanegas *et al.* 1974; Schmidt 1979). We found four groups in the perch (Guthrie and Banks, unpubl. obs.).

One of the tests applied by many researchers is to stimulate the eye with a weak flash of light prolonged for several seconds. Purely excitatory or inhibitory receptive fields produce 'on' or 'off' responses in the fibre (i.e. impulses follow the 'on' or 'off' of the flash). If there are overlapping excitatory and inhibitory fields (opponence) as is frequently the case, then both 'on' and 'off' responses occur. In the goldfish 'on' fibres were slightly less common than 'off' fibres (Ormond 1974; Schellart and Spekreijse 1976), but in the perch they were less than half as common. In goldfish Ormond found 'on-off' fibres more abundant than either 'on' or 'off' combined, but in the perch we found about as many as there were 'off' fibres alone. These observations accord reasonably well with the idea of an upward-looking predator like the perch needing fewer 'on' fibres (light object on dark), because most objects are seen in silhouette, than a bottom-feeding carp or goldfish would need, which sees objects against a darker ground.

Wartzok and Marks (1973) were able to measure the receptive field diameters of retinal ganglion cells in the goldfish, and found them to be mostly 3—12° across (angle subtended at the eye). To our great surprise we found that similar methods to those used by these authors, and that we had applied successfully to perch tectal cells, failed to excite perch retinal fibres. Only seven out of a sample of 110 fibres provided us with a clearly defined receptive field. The prolonged tonic responsiveness of tectal cells which allows the mapping technique to be used so easily seemed to be absent from most perch efferents encountered. This lack of sensitivity and rapidity of adaptation can be partly countered by using very bright brief flashes;

to which 70 per cent of fibres will respond. Wartzok and Marks provided convincing evidence not only for directional sensitivity in their goldfish fibres, but also for considerable differentiation of responses to the velocity of a stimulus spot. There were indications of velocity-tuned units, which could provide a basis for discrimination of rates of prey movement, etc.

A final question is how many of the optic nerve fibres are efferents? Sandeman and Rosenthal (1974) showed that the proximal stump of the optic nerve of the trigger fish contained fibres that responded to vestibular, tectal and other stimuli. Vanegas (1975) observed that the fastest massed potential wave that could be recorded from the optic nerve of *Eugerres* as a result of electrical stimulation produced action potentials in tectal cells without a synaptic delay, and with no evidence of fatigue at high frequencies. This could only occur if the axon being stimulated was part of the cell being recorded from, i.e. there was no synapse between them, and the axon was a retinal efferent. Studies on both fish and birds (Sandeman and Rosenthal 1974; Miles 1972) point to a facilitatory role for retinal efferents, but Banks and Guthrie (forthcoming) found heightened excitability of the perch retina to a double light flash of high intensity followed partial tectal ablation (Figure 8.6). Note the disappearance of the weak slow wave and the early appearance of the second response. These suggest an inhibitory function for the tectum.

The Visual Functions of the Optic Tectum

Histological examination of the tectal lobe of a fish in which the corresponding eye has degenerated, or been surgically removed, reveals a marked reduction of two of the outer layers, SO, SFGS (see Vanegas 1975 for terminology), whilst the deeper layers, SGC, SAC, PVC, are seen to be more or less normal. These outer layers are the ones in which most of the retinal afferents (axons of ganglion cells) terminate, and recordings from them reveal a fairly precise relationship between the position of a point of light in the visual field of the eye and the position of excited fibre terminals in the tectum. These matching positions show that the spatial relationships of visual stimuli are preserved in the pattern of excited afferents, and recordings from intrinsic tectal cells show that many of these conform to this pattern as well. One of the major functions of the outer part of the tectum concerns object position, a vital function both for the localisation of food, and for obstacle avoidance.

Deeper lying tectal cells are rather less likely to conform precisely to the retinotopic map in the position of their receptive fields. They often have rather large receptive fields which overlap more with others, and are less easily excited by visual stimuli alone (Ormond 1974; Guthrie and Banks 1978). Further, their responses are often more complex from the point of view of facilitation and habituation (Vanegas, Amat and Essayag-Millan 1974). The degree of stratification is much less striking than it is in frogs and toads, probably due to the much longer dendrites of fish tectal neurons mentioned earlier. Recent work by Niida *et al.* (1980) has proved cautionary in that one morphological type (type I, pyramidal cell) has been shown by him to be capable of quite different kinds of response.

There is some differentiation of the receptive field shapes of intrinsic tectal cells at least in the perch (Guthrie 1981; Guthrie and Banks 1978), and this would support the idea that a degree of spatial discrimination exists at the tectal level. However, it may originate at the retinal, rather than the tectal stage, and Schellart and Spekreijse (1976) have proposed that only the very complex multicentre fields represent tectal processing. In the perch these are well represented, and there are also cells which appear to be specialised to respond to moving bars irrespective of the angle at which they are moved across the receptive field centre (Guthrie 1981).

Double-opponent colour coded receptive fields are characteristic of a high percentage (68 per cent) of goldfish afferents (Daw 1968), but we found them to be less common (18 per cent) amongst tectal cells in the perch (Guthrie 1981). Of particular interest to us were cells responding maximally to wavelengths in the 680—700 nm band, which might be important in the discrimination of the vermilion fin patches characteristic of perch. Despite the absence of a blue-sensitive cone in the perch and the masking effect of corneal pigment, some cells show quite a marked sensitivity to the short wavelengths.

Many years ago Buser showed that the tectal evoked response in the catfish contained wavelength-sensitive components (Buser 1955) when stimulated by light flash. This work was followed up by Konishi (1960) using goldfish, and he obtained even more striking results. The early red-sensitive wave would disappear altogether if the eye was stimulated by blue light. We have examined this phenomenon in the perch (Figure 8.7) where it is very much weaker. Figure 8.8 illustrates the difference in the amplitude of the waves composing the TER when the wavelength of the light stimulus is varied. Note the large amplitude of all major components at the long wavelengths.

Figure 8.7: The tectal-evoked response to a flash of white light in the perch recorded with wire electrodes. Negativity upwards. Note the very well defined potential waves. Terminology after Konishi (1960).

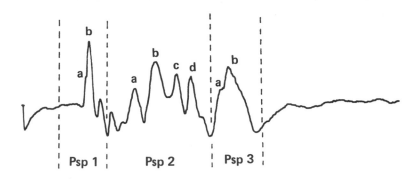

There are 8 red-sensitive cones to each green-sensitive cone in the perch. One interesting feature was a wave (Psp2b) that only appeared to be distinct when the stimulus was at 535 and 563 nm (green, blue-green). A link between latency and wavelength may be without any significance beyond indicating the existence of biased channels within the retinotectal pathway. This follows from the predominance of red 'on' centre types of receptive field over other types in the goldfish (Daw 1968), which we believe also exist in the perch. It does, however, raise the possibility of wavelength differences being coded as latency differences, where the eye of a fish was being displaced across a colour pattern. The latency differences observed by Daw (1968) between red 'on' centre responses and those from green 'on' surrounds were so large that such coding might be reasonably resistant to the effects of intensity differences. The latter would tend to be smallest at boundaries between coloured areas.

Summary

In this brief survey of some of the aspects of teleost neurobiology that are of interest at the present time, I have omitted much that would seem important to specialists in other fields than my own. I hope they will forgive me for this. Over the 350 million years that have elapsed since fish first appeared in any numbers in the fossil record, they have

Figure 8.8: Differential responses of the Psp1 (●———●) and Psp2 (○---○) components of the tectal-evoked responses (perch tectum; see Figure 8.7) to light of different wavelengths. All points computer averaged from 16 responses.

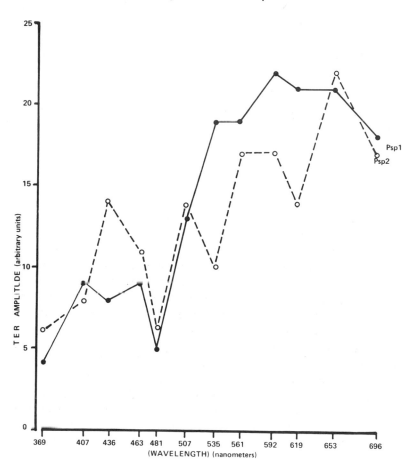

diversified into 395 modern families, and live in every kind of aquatic habitat. Part of their success has been due to the development of refined sensory abilities, and a basic design of central nervous system capable of effective sensory processing. An understanding of the way in which integrative functions are carried out in a vertebrate brain with a comparatively small complement of interneurons will, I believe, give us special insights into control systems generally.

Acknowledgements

The work on perch that I have mentioned here has been carried out with the help of a number of researchers in the Fish Laboratory at Manchester. I would like particularly to thank J. R. Banks, my collaborator on the perch studies, and Mrs Veronica Frater for preparing the manuscript for publication.

9 HORMONES AND AGONISTIC BEHAVIOUR IN TELEOSTS

A.D. Munro and T.J. Pitcher

Specific areas of the brain are targets for circulating hormones in birds and mammals (Leshner 1978; Pfaff 1980), and although the anatomical, physiological, biochemical and behavioural evidence is not as extensive, a similar situation occurs in teleost fish. Hormone-sensitive neurons in some brain areas exercise feedback control on the pituitary gland whilst other areas control the expression of behaviour. Furthermore, in higher vertebrates it is known that performance of particular behaviours may itself alter endocrine function (Leshner 1981). Slater (1978) gives an introductory review of work on birds and mammals.

Hormones may exert a temporary or a permanent alteration in behavioural responsiveness. For example, castrated adult male mice show little intermale aggression, but normal levels of aggression are restored by injections of androgen: a short-term temporary action. Androgen injections into adult female mice, on the other hand, have no such result, unless the females were injected with androgen as neonates: an example of a permanent alteration where the presence of a hormone at a particular stage of development 'primes' the central nervous system so that certain stimulus situations may evoke a particular response in later life (Leshner 1978).

The aim of this chapter is to review our knowledge of the inter-relationships between the endocrine system and the expression of agonistic behaviour in teleosts. Agonistic behaviour covers all the elements of behaviour seen during aggressive encounters including submission and fleeing.

Endocrine Targets in the Teleost Brain

Extracts of teleost brain from most regions contain receptors with a high affinity for oestradiol (Myers and Avila 1980). Oestrogen can also be synthesised by fish brain extracts since brain cells contain an aromatase capable of converting exogenous testosterone and methyltestosterone (Callard et al. 1981).

The regional distribution of sex steroid receptors has been demonstrated by autoradiography in the brains of four teleosts: goldfish (*Carassius*), swordtails (*Xiphophorus*) (both Kim, Stumpf, Sar and Martinez-Vargas 1978), sunfish (*Lepomis*) (Demski and Hornby 1982), and paradise fish (*Macropodus*) (Morrell and Pfaff 1978). Steroid and thyroid hormones probably exert their effects on nervous tissue by influencing the transcription of DNA in the cell nucleus (McEwan, David, Parsons and Pfaff 1979). Neurons which are steroid-receptive are therefore defined as those autoradiographically shown to concentrate radioactively-labelled sex steroid within their cell nuclei. Such cells have been identified only in the forebrain of teleosts (unlike tetrapods: Morrell and Pfaff 1978) and the overall distribution is similar for both sexes (where tested) and for testosterone and oestradiol. The contribution of aromatase to the exact overlap is unknown, nor is it clear whether the same neurons bind both steroids.

The nomenclature of the various cell groups ('nuclei') in the teleost forebrain has been rationalised in a review by Munro and Dodd (1983) and discussion in this chapter will follow the new system.

Within the anterior subdivision of the forebrain (telencephalon), neurons in the anterior ventral (Vv) and the supracommissural (Vs) regions sequester steroids in their nuclei in three teleosts studied, but not in sunfish (Figure 9.1). The various cell fields of the dorsal telencephalon do not appear to become radioactively labelled, and this conflicts with biochemical studies on a cichlid (*Hemichromis*) (Myers and Avila 1980), and with undoubted steroid-sensitive neurons in the pallial homologue of this region in tetrapods (Kim *et al.* 1978). The disparity, which needs resolution, may be a consequence of species differences, insensitivity of the autoradiographic technique, or it could result from a cytoplasmic, non-genomic action of oestradiol in the teleost dorsal telecephalon (Kelly, Moss and Dudley 1976).

In the teleost diencephalon (Figure 9.1), cell nuclei of scattered neurons in the thalamus may be labelled (Kim *et al.* 1978), but the major target is the hypothalmus. Oestrogen is taken up by nuclei of cells in the preoptic area, including some of those in the preoptic nucleus, and by the posterior region of the lateral tuberal nucleus. In paradise fish and swordtails a third steroid-sensitive field has been reported in the lateral recess nucleus and in the monoaminergic posterior periventricular nucleus (Figure 9.1).

There is also electrophysiological evidence for hormone action on

Figure 9.1: (below) Diagrammatic distribution of some of the
major midline cell masses of the forebrain in a typical teleost
(nomenclature based on Munro and Dodd 1983). A—A (over page)
Cross-section of the telencephalon and hypothalamic preoptic
area at the level of the line A—A. B—B Cross-section of the
diencephalon and adjacent midbrain at the level of the line B—B.
The distribution of sex steroid-concentrating neurons is indicated
by the dotted areas, although each of these areas need not show
such a property in all teleosts (see text; Figure 9.5). Figure 9.1
modified from Kim *et al.* 1978. AC, anterior commissure; ADT,
dorsal telencephalon (pallial homologue); AVT, ventral
telencephalon (subpallium); DThal, dorsal thalamic nuclei; GHab,
habenular ganglion; nLT, lateral tuberal nucleus; nPA, anterior
periventricular nucleus; nPP, posterior periventricular nucleus;
nPrG, posterior portions of the preglomerulosus complex; nRL,
nucleus of the lateral recess; nRP, nucleus of the posterior recess;
OB, olfactory bulb; OC, optic chiasma; Pit, pituitary; POA, preoptic
area; RL, lateral recess of the third ventricle; Vs, supracommissural portion of the AVT; VThal, Ventral thalamic nuclei;
Vv, ventral portion of the AVT.

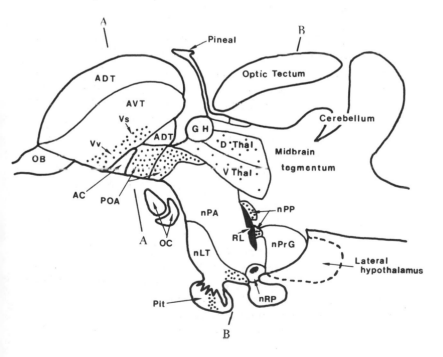

Figure 9.1 continued

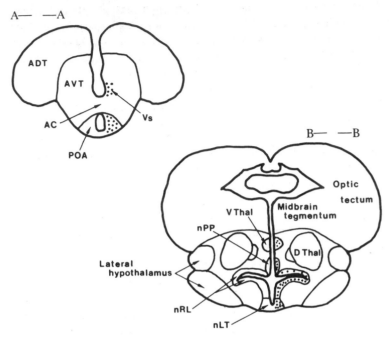

the fish brain, in addition to the anatomical and biochemical work outlined above. Gorbman and his co-workers have demonstrated that exogenous sex steroids and thyroid hormones can modify the processing of olfactory and visual information in goldfish. As much as nine days prior treatment may be required for a hormone to have an effect (Hara, Ueda and Gorbman 1965).

The optic tectum in the midbrain is the principal target for retinal projections in teleosts (Chapter 8). Thyroxine appears to sensitise the tectum to stimulation of the retina with light, whereas neither oestradiol nor testosterone had any effect (Hara *et al.* 1965). However, high doses of steroids were used, which also failed to affect olfactory activity.

In three-spined sticklebacks (*Gasterosteus*), females are more sensitive to red light than males in summer, but not in winter (optomotor test: Cronly-Dillon and Sharma 1968), suggesting direct tectal steroid sensitisation. However, neurons in the dorsal thalamus also receive a retinal input and project to the telencephalon and optic tectum (Munro and Dodd 1983). Thalamic neurons which bind radioactive steroids may feature in these interconnections and could modify visual responsiveness.

Thyroxine and sex steroids modify olfactory activity in the goldfish brain, which is interesting given the drastic modifications of male behaviour in response to female pheromones at spawning (Partridge, Liley and Stacey 1976). Sensory cells in the olfactory epithelium, whose sensitivity is enhanced by oestradiol (Hara 1967), send fibres to the glomerular layer of the olfactory bulb (Figure 9.1) where analysis occurs (Munro and Dodd 1983). Neurons in the bulb relay processed information to the telencephalon and elsewhere. Activity of these bulbar neurons is enhanced by thyroxine and by inter-mediate, but not high, doses of oestradiol and progesterone (Oshima and Gorbmann, 1966a,b, 1968, 1969).

Where thyroxine may exert its effect is not clear: as with sex steroids, a direct effect on the sensory epithelium cannot be ruled out. For sex steroids, there is no autoradiographic evidence for uptake by bulbar neurons in three of the teleosts studied, although information is not available for goldfish. The telencephalic field Vs (Figure 9.1) includes neurons projecting back to the olfactory bulb (Munro and Dodd 1983) and this centrifugal pathway may modulate bulbar activity. Electrophysiologically, the olfactory bulb receives both excitatory and inhibitory influences from the rest of the brain, the axons of the Vs neurons possibly representing a common final pathway. One excitatory input to the bulb, from an unknown brain area, is depressed by thyroxine and another by progesterone, which acts on part of the forebrain. Intermediate doses of oestradiol, by way of the midbrain or below, act similarly (Oshima and Gorbman 1966a, 1968, 1969). These workers also found that progesterone acting below the midbrain facilitates bulbar activity. Hara and Gorbman (1967) found the same for testosterone in male, but not female, goldfish. By an action on neurons in the Vs or preoptic area, testosterone may be responsible for male sensitivity to female spawning pheromone (Hara and Gorbman 1967; Demski and Hornby 1982).

The conflict between autoradiographic and electrophysiological evidence concerning the action of oestrogen on midbrain or lower levels in goldfish suggests that not all sites of oestradial action have been located by autoradiography (Myers and Avila 1980).

Agonistic Behaviour and its Functional Significance

Agonistic behaviour forms a complex of attack, threat, submissive and fleeing action patterns, usually directed towards a conspecific, and is considered to arise from conflicting motivations to attack and withdraw (Hinde 1970, 1982), although some displays may be ritualised during evolution, especially in courtship (Baerands 1975). Predatory behaviour is distinct ethologically (Huntingford 1976) and pharmacologically (Figler 1981) in teleosts, as in mammals. Scott (1981) has argued that agonistic behaviour may have evolved from antipredator behaviour, and in sticklebacks there is weak evidence for a shared motivation (Huntingford 1976, 1982).

Regardless of the origins or motivation of agonistic behaviour, its components are difficult to categorise except as a complex of defined action patterns. Rather than consider this whole complex, Rowell (1974), Barlow and Ballin (1976), and Miller (1978) have pointed out that many workers on teleosts, as on mammals, have considered only the 'aggressive' elements, thereby presenting incomplete, and perhaps misleading, views of social relationships. Aggressive behaviour itself proves almost impossible to define generally (Hinde 1970).

Agonistic behaviour occurs during two types of conflict situation: self-defence (which may include antipredator behaviour) and property- or resource-defence (McFarland 1981). Defended resources, often a territory, include shelter (salmonids: Jenkins 1969), localised food supply (medaka: Magnusson 1962), spawning site (minnows: Pitcher, Kennedy and Wirjoatmodjo 1979), progeny (cyprinodonts: Kodric-Brown, 1978), or even mates, progeny and feeding areas in a complex territory (cichlids: Taborsky and Limberger 1981).

At low population density, cichlids are territorial in aquaria (Vodegel 1978) but as density increases, a smaller proportion of individuals establish territories until at high density a dominance hierarchy tends to be formed. Although sometimes dismissed as an artifact (Rowell 1974), such hierarchies have been reported for crowded groups of coral reef teleosts (Loiselle and Barlow; in Miller 1978) in the wild. Similarly, Jones (1983) observed that a wrasse (*Pseudolabrus*) was territorial at field sites with a low to medium population density, but shoaling at very high density sites. At the end of the dry season, cichlids may encounter high densities and so their behaviour in the aquarium should not necessarily be regarded as aberrant. Nevertheless, it is important to take care when extrapolating from aquarium studies or endocrine experiments to agonistic behaviour in the wild.

The Influence of Hormones on Agonistic Behaviour in Teleosts

Hormones of the pituitary—gonad and pituitary—adrenocortical axes are of principal importance in the control of agonistic behaviour in birds and mammals, although there is considerable intraspecific variability (Leshner 1978, 1981). Teleosts in general have been studied less rigorously: work on several families was described by Liley (1969) and work in five of these is discussed here. In nearly all experiments, only the attack (= aggressive) responses and displays have been recorded to the exclusion of the other components of agonistic behaviour.

The Pituitary—Gonad Axis

The Three-spined Stickleback.   Lengthening days and rising temperatures in spring stimulate the onset of the breeding season in *Gasterosteus aculeatus*, whose reproductive biology has been reviewed by Wootton (1976). Secretion of testicular androgen stimulates the male's secondary characters at this time, which include red belly, blue eyes and a kidney modified to produce glue for sticking nest material together.

Whilst females shoal throughout the year, males become more aggressive in spring as the gonads develop. This initial rise in non-territorial aggression (Figure 9.2) is not blocked by castration or methyltestosterone treatment (Hoar 1962; Wai and Hoar 1963; Baggerman 1966; Wootton 1970), but can be inhibited by short photoperiod and induced in immature fish by mammalian gonadotrophin, but not by other pituitary preparations. Hoar (1962) therefore suggested that the early aggressive phase was caused by pituitary gonadotrophin (GTH) secretion independent of testicular hormones, although Baggerman (1968) has questioned this. GTH secretion is affected by the feedback of aromatisable androgen (Crim *et al.* 1981; Chapter 11). Depending on whether a positive or a negative feedback mechanism is operative at this stage in the male stickleback, GTH secretion under a long photoperiod would be increased by either methyltestosterone or castration. A critical test for Hoar's hypothesis concerning the early aggressive phase would be the transfer of previously castrated fish from short to long days, with and without androgen replacement therapy with simultaneous studies of GTH production.

Nest building, courtship and parental behaviours are abolished by castration or anti-androgen treatment, these effects being reversed by methyltestosterone (Smith and Hoar 1967; Baggerman 1968; Rouse, Coppenger and Barnes 1977). Nest building is associated with territory formation and a dramatic increase in aggression. This may be reversed by castration, although the experiments give inconsistent results, possibly because of differences between the wholly freshwater *G. leiurus* stickleback and the anadromous *G. trachurus* (Baggerman 1966; Wootton 1970). Territory defence may not be controlled by androgens directly, but rather by the immediate visual stimulus and associated 'ownership' behaviour of the con-structed nest (Wootton 1970).

Agonistic encounters have been reported between females in the aquarium, but not in the wild (Wootton 1976; Huntingford 1979). Females lack the threat displays of males, which suggests that androgens have an early developmental influence. Injections of mammalian gonadotrophin with methyltestosterone stimulate the construction of an imperfect nest and territoriality in 6 per cent of females, suggesting perhaps a residual responsiveness of the adult female brain (Wai and Hoar 1963) but courtship behaviour was said not to be exhibited. Any effect of GTH on adult female behaviour may be over-ridden by ovarian secretions: aquarium females become less aggressive as their ovaries mature (Sevenster and Goyens 1975; Huntingford 1979), this being reversed by ovariectomy (Wai and Hoar 1963; Liley 1969).

Figure 9.3 illustrates a model for the endocrine control of agonistic behaviour in sticklebacks based on these observations.

Centrarchid Sunfish.   In spring males leave mixed-sex shoals to establish territories and build nests. Increases in day length and temperature appear to trigger this behaviour (Smith 1969), and its endocrine control has been studied in two species, *Lepomis gibbosus* and *L. megalotis* (Smith 1970; Fiedler 1974).

Gonadal androgens seem to control nest building (Smith 1969). Fiedler (1974) considered that methyltestosterone-stimulated aggression, but Smith (1970) argued that this was an indirect effect similar to that advocated for the stickleback by Wootton (1970). Artificial crowding of sunfish stimulates aggression irrespective of

Figure 9.2: Schematic summary of the effects of endocrine manipulations on the relative frequency of aggressive and associated behaviours in the male three-spined stickleback, plotted against a standardised time scale. (a) Aggression and sexual behaviour of males kept on a short daylength. (b) Aggression and sexual behaviour of males transferred to a long day cycle. (c) Nest building and associated care by males transferred to long days. (d) Aggression and sexual behaviour of males, if transferred back to short daylengths prior to nest-building; or sexual behaviour of castrated fish maintained under long day lengths. (e) Aggression of males castrated prior to nest-building and maintained under long day lengths. (f) Aggression and sexual behaviour of males (*G. trachurus*) castrated about 14 days after the start of nest-building. (g) Aggression in males (*G. leiurus*) castrated an unstated time after the start of nest-building; aggression against females wanes more slowly than that against males.

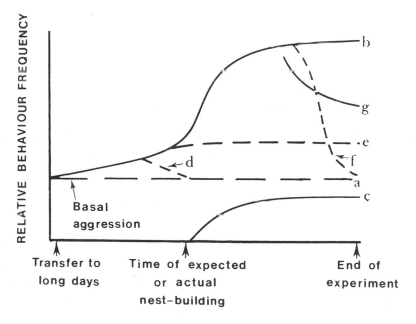

a, b, c, d, e, f, based on Baggerman (1966); d, e, also based on Hoar (1962); and g based on Wootton (1970)

Figure 9.3: Summary of the possible mechanisms by which the pituitary—gonadal axis is considered to modulate aggressive behaviour in the male three-spined stickleback, as described in the text. As indicated in this diagram, the act of building a nest may mediate the effects of gonadal androgen on aggression, and also may modify the activity of the pituitary—gonadal axis itself. +, likely facilitatory action; (+), possible facilitatory action; +/—, likely facilitatory and/or inhibitory actions.

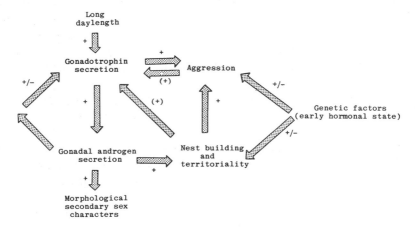

sex or gonadal state and since such crowding occurs naturally on the spawning leks of the males, agonistic behaviour would be expected to increase in any case.

Anabantoid Labyrinth Fish.   Several species of Anabantoidei have been studied: paradise fish (*Macropodus opercularis*), Siamese fighting fish (*Betta splendens*), the blue gourami (*Trichogaster trichopterus*) and the dwarf gourami (*Colisa lalia*).

Isolated individuals of paradise fish of both sexes make equal numbers of attacks on mirror images, but males show significantly higher levels of lateral display from the time of testis maturation (Davis and Kassel 1975). Methallibure abolishes this sex-difference, possibly due to inhibition of the pituitary—gonad axis (Davis, Mitchell and Dolson 1975). Methyltestosterone is variously reported as enhancing aggression (Fiedler 1974) or having no effect (Davis 1975), the disparity perhaps being due to social isolation which increases attack behaviours and may mask the androgen's effects (Davis 1975).

Weis and Coughlin (1979) reported that castration did not affect frequency of gill-cover erections in male fighting fish. Testosterone may, however, increase the relative frequency of lateral displays (Bevan and Munro, unpubl. obs.), for similar reasons to those in paradise fish.

In the blue gourami, castration does not affect attack behaviours (Johns and Liley 1969). Without giving details, Forselius (1957) reported that methyltestosterone increased aggression and activity in both sexes of dwarf gourami, whilst oestrone reduced male aggression.

The evidence for gonad involvement in agonistic behaviour in anabantoids is therefore confusing, perhaps because of differences in methodology and the behaviours scored.

Bathygobius soporator (Gobiidae).   The males of this marine goby become territorial during the breeding season, chasing off all conspecifics except mature females, which are courted (Tavolga 1955). Besides abolishing the male's secondary morphological sexual characters, castration also eliminates attack behaviour so that all intruders are courted. This applies even to fish of other species, and suggests that a testicular secretion directly controls attack, but not courtship (Tavolga 1955). Hypophysectomy abolishes courtship as well as agonistic behaviour, and reduces activity. A crucial test would be to try replacement therapy with androgens in juveniles, males and females (Baggerman 1968).

Cichlid Fish.   This family of perciforms, which includes 10 per cent of all known teleost species, exhibits two principal forms of reproduction (Fryer and Iles 1972). First, in the substrate-spawners pairs of fish establish a spawning territory in the wild, wherein to raise the fry. Relatively little work has been done on the endocrine control of behaviour in this type. In the second, apparently derivative strategy of mouthbrooding, it is typically only the male which stakes out and guards a spawning territory, or lek, from which he courts females. After spawning, the female picks up the eggs in her mouth, leaves his territory and sometimes herself becomes territorial (Vodegel 1978).

Ovariectomy of a substrate-spawning cichlid, the blue acara (*Aequidens pulcher*), eliminates territorial behaviour (Munro 1982). For hierarchical groups of maturing females, testosterone increases and oestradiol depresses agonistic encounters (Munro and Pitcher 1983), while activity is unaffected. Hormones were administered by an immersion technique; testosterone was less effective at 250 µg/l than at double that concentration, suggesting a dose-dependent effect. The clearly contrasting effects of testosterone and oestradiol on behaviour show that the former was not aromatised prior to its action.

Maturing female blue acara, when isolated and presented with models of conspecifics or mirrors, give comparable results to the hierarchical groups for the effects of testosterone and oestradiol, a result which supports the hypothesis that these hormones act centrally to alter agonistic behaviour (Munro and Pitcher 1983). Progesterone may also stimulate aggressive behaviour in isolated fish (Munro and Pitcher, unpubl. obs.), although this could be an indirect effect by way of gonadal metabolism to form an androgen. A possible direct effect of GTH on the blue acara remains to be tested.

In the closely related brown acara (*A. portalegrensis*), recognisable agonistic behaviours (chasing, fleeing, approach and avoid) appear by 30 days after hatching (at 28°C). This behaviour appears about two weeks before steroidogenic tissue can be recognised in the gonads (Munro 1982). This suggests that agonistic behaviour is not totally dependent upon endogenous sex steroids.

Castration of Egyptian mouthbrooders (*Pseudocrenilabrus multicolor*) abolishes territoriality, an effect reversed by injections of testosterone, which also induces territory defence in females (Rixner, cited by Reinboth 1972). In another mouthbrooder (*Haplochromis burtoni*), testosterone increases attack frequency by territorial adult males on small blinded aggressee fish (Fernald 1976), and on non-territorial conspecifics (Wapler-Leong and Reinboth 1974), but, in contrast to our findings on the blue acara, not on fish models. Testosterone-induced territory defence in female *H. burtoni*, which Wapler-Leong and Reinboth also found, did not react to fish models. The responsiveness of *H. burtoni* to various models has received considerable attention, and the negative effects of testosterone on this suggest a subtle hormone action (Heiligenberg 1974; Wapler-Leong and Reinboth 1974).

Figure 9.4 summarises the known effects of sex steroids on agonistic behaviour in cichlids.

Problems of Methodology in Experiments on Sex Hormones and Behaviour in Teleosts

Whilst the pituitary—gonad axis modifies agonistic (aggressive) behaviours in at least some teleosts, the hormones involved appear to

Figure 9.4: Summary of some of the endocrine axes which are postulated in the text to modulate agonistic behaviour in substrate-spawning cichlids. No attempt has been made to include likely feedback mechanisms. ACTH, adrenocorticotrophin; E$_2$, oestradiol; F, cortisol; GTH(s), gonadotrophin(s); P, progesterone; T, testosterone; T$_4$, thyroxine; TSH, thyroid-stimulating hormone; —, likely inhibitory action; (—), possible inhibitory action; for other abbreviations see Figure 9.3.

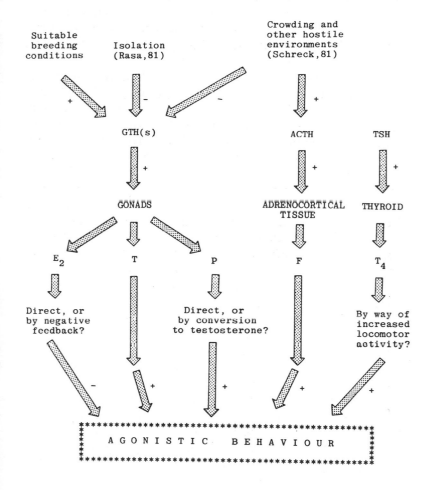

vary both within and between species. Is this the result of real biological differences, or may it be due in part to the different techniques
of experiments?

In order to validate a role for GTH in the control of aggression,
future experiments should use gonadectomised, and preferably also
hypophysectomised, fish.

Gonadectomy studies must be interpreted with caution, as
undetectable fragments of steroidogenic tissue may remain (Johns
and Liley 1969). Also, the adrenocortical tissue is an additional
source of testosterone. Direct measurements of plasma sex steroids
are clearly a necessary adjunct of gonadectomy studies. A similar
need for GTH measurements has been pointed out above for sticklebacks.

In the face of negative results for one gonadal steroid, a short-term
gonadal influence on aggression need not be eliminated; too high or
too low a dose may be ineffective (Leshner 1978). Unlike testosterone, methyltestosterone is ineffective in mice, despite being
aromatisable (Floody and Pfaff 1974). The behavioural effects of the
principal teleost androgen, 11-ketotestosterone (Chapter 11),
require testing. Like aromatisable androgens, oestrogens feedback
on pituitary GTH secretion (Crim *et al.* 1981; Munro and Dodd
1983) and indirectly may reduce aggression in the blue acara by
depressing androgen production.

To avoid the discontinuous and traumatic features of injection
protocols, immersion or implant techniques would appear to be the
methods of choice. Where necessary, assays can be used to monitor
circulating hormone levels. An adequate period of exposure must be
allowed.

Protocols for measuring agonistic behaviour must allow for the
wide variation in responsiveness between individuals, and increases
which are merely due to changed activity levels or encounter rates
must be sought in data. Differences between local stocks of some
species of fish may affect results (Huntingford 1982). Prolonged
isolation may enhance aggressiveness in many teleosts (Davis 1975),
except in cichlids where the reverse is true (Heiligenberg 1974; Rasa
1981), and these processes may mask hormonal influences in an
experiment. Treatments may alter the activity of other endocrine
systems or the secondary sex characters of test fish in a group, indirect
effects which should be excluded as far as possible.

As wide a range of action patterns as possible should be recorded
in the behavioural experiments, to minimise the problems of

arbitrarily labelling certain behaviours 'aggressive'. Multivariate analyses should then be employed to identify the categories of agonistic behaviour which respond to the hormone treatments (Huntingford 1976; Munro and Pitcher 1983).

The Role of the Pituitary—Gonad Axis in Agonistic Behaviour

Allowing for the criticisms of the previous section, two basic classes of teleosts can be identified from the experiments outlined above. The pituitary—gonad axis modulates agonistic behaviour in stickle-backs and cichlids in different ways.

Male sticklebacks build and defend a nest, and show brood care: the usual strategy of parental care to evolve in fish without internal fertilisation (Maynard-Smith 1977). These behaviours are sex-specific, as might be expected, and are presumably genetically determined by sex steroid priming. The available evidence suggests that sunfish and anabantoids are similar, to the extent also that courtship is abolished by castration (Liley 1969).

In substrate-spawning cichlids, both sexes engage in territory defence and brood care, whilst in the moutbrooders the sexual roles tend to be partitioned. Here there is no evidence for sexual dimorph-isms in the response of either cichlid reproductive guild to exogenous testosterone.

Bathygobius utilises a breeding strategy similar to the stickleback. However, like cichlids, evidence suggests that aggression is testes-dependent, but courtship is not (Liley 1969). It is not clear why *Bathygobius* should occupy this intermediate position.

In female salmonids (Chapter 11), plasma androgens peak just prior to ovulation and progestagens just after this. In female substrate-spawning cichlids, a similar cycle would produce pre-spawning and parental aggression as observed in the wild: the hormonal control of parental behaviour will be returned to below.

Sites and Mechanisms of Steroid Action

Aromatisable androgens and oestrogens probably control pituitary GTH secretion partly be feedback on the lateral tuberal area (Munro and Dodd 1982). By thereby reducing gonadal androgen secretion, oestradiol may depress agonistic behaviour in the acara (Munro and Pitcher 1983). Oestradiol may regulate prolactin secretion via neurons in the posterior periventricular nucleus (Munro, in prep.).

Hormones may act at several stages in the neural control of an animal's behaviour (Leshner 1978; Pfaff 1980). They may alter the reception and processing of sensory information, the control of motor outputs for specific action patterns or intervening stages such as motivational tendencies.

The failure of testosterone to elevate attack on dummies in *Haplochromis* may be an example of the first mechanism, as is the sensitivity of male goldfish to female spawning pheromone. Sex differences for displays in sticklebacks and anabantoids may be an example of the second.

The control of general behaviour is by the telencephalon regulating lower brain centres. This is mainly by the inhibition of behaviours receiving no positive reinforcement, so that previously dominant behaviours persist after telencephalon ablation (Flood and Overmeir 1979). Such an operation induces 'fearfulness' (Huntingford 1976), fish being less aggressive (de Bruin 1980).

In cichlids, testosterone may be a positive reinforcer for aggression. Unfortunately, steroid-binding neurons have not yet been mapped in the cichlid brain, although there is biochemical evidence for oestrogen receptors in both the dorsal (ADT) and ventral (AVT) telencephalon (Myers and Avila 1980). In other teleosts, both excitatory and inhibitory loci have been identified in the ADT (McIntyre and Healy 1979; de Bruin 1980), but classical steroid-sensitive neurons are described only in portions of the AVT (Figure 9.1). Unfortunately, sex steroids may have only an indirect effect, if any, on aggression in those species studied with autoradiography (Liley 1969, and above).

Therefore, in cichlids, sex steroids may modulate aggression by acting on an unknown group of neurons in the ADT, or else on neurons described in other species as steroid-sensitive, either in the AVT, or in the hypothalamus. Figure 9.5 summarises the evidence of stimulation, lesioning and autoradiographic studies in relation to teleost aggression. It will be apparent that electrical stimulation and lesioning of certain steroid-sensitive hypothalamic areas may modify aggression. They are also related by descending fibre connections. The precise importance of sex-steroid action on these various prospective targets must await hormone-implant and lesioning studies in cichlids, together with autoradiography. The possibility of a non-genomic steroid action on other centres must be considered (Myers and Avila 1980).

Most teleosts show little reproductive aggression and no brood care. Families which do show such behaviours have evolved independently. Such convergent evolution of reproductive strategies

Figure 9.5: Summary of the possible levels at which androgens (and oestrogens?) may act to modulate agonistic behaviour in cichlids. The major descending axonal pathways (see Munro and Dodd 1983) are indicated: the open arrows represent the possible paths involved in the expression of agonistic behaviour. LFB, lateral forebrain bundle; MFB, medial forebrain bundle; *, areas possessing steroid-concentrating neurons (Figure 9.1); x, areas where electrical stimulation results in evoked aggression (Demski and Knigge 1971; Demski 1973); +, areas whose destruction facilitates or inhibits aggression (McIntyre and Healy 1979; de Bruin 1980); other abbreviations as in Figure 9.1.

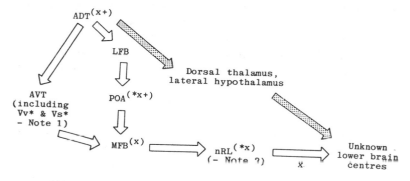

Note 1: Hormone-sensitive neurons have not been described in the AVT of *Lepomis* (Demski and Hornby 1982). Note 2: Steroid-sensitive neurons have not been described in the nRL of either *Carassius* or *Lepomis* (Kim *et al.* 1978; Demski and Hornby 1982)

may, or may not, be due to the alteration of the same basic neural pathways: in this light, the effects of hormones on typically shoaling or schooling (Pitcher 1979) teleosts are of considerable interest.

The Pituitary—Adrenal Cortex Axis

In mammals, acute treatment with ACTH appears to increase aggression indirectly through stimulation of adrenal glucocorticoid secretion. In contrast, chronic treatment with ACTH reduces aggression directly and increases submissiveness indirectly by increasing glucocorticoid secretion (Leshner 1978, 1981).

Various stresses stimulate glucocorticoid secretion in teleosts as in mammals, including low rank in a dominance hierarchy (Schreck 1981). Recent work on mammals suggests that increased gluco-corticoid secretion is a cause as well as an effect of maintained low social status in a dynamic feedback loop (Rowell 1974; Leshner 1981).

ACTH may facilitate memory in fish as in mammals (Borsook, Woolf, Vellet, Abramson and Shapiro 1981). It has been suggested to induce arousal in damselfish (Rasa 1971) and to stimulate aggression in both sexes of a tilapia, *Oreochromis mossambicus* (Fiedler 1974). All these effects were considered to be direct, independent of the adrenal cortex.

Cortisol, the principal teleost glucocorticoid, increases attacks on fish models and mirrors by blue acara, and increases agonistic behaviours in hierarchical groups. The pattern of agonistic behaviours in cortisol-treated hierarchies differs from that seen in those treated with testosterone, suggesting a different functional role and action on the brain (Munro and Pitcher 1983).

Some heavy metals stimulate cortisol secretion (Schreck 1981). Sublethal doses of such pollutants may stimulate aggression in groups of sunfish (Henry and Atchinson 1979). Various stressors therefore may enhance agonistic behaviour in social groups of teleosts by stimulating the adrenocortical—pituitary axis. In the wild, dispersal may result, but in laboratory conditions high levels of aggression occur in such experiments.

On the basis of research in these fields to date, we speculate that testosterone is the hormone of resource-defence whereas cortisol is that of self-defence (Figure 9.4). Although this may appear to be an oversimplication (Munro and Pitcher 1983), behavioural experiments might be designed to test the idea.

The Pituitary—Thyroid Axis

Thyroid hormones seem to increase locomotor and metabolic activity in teleosts and have a role in migration (Liley 1969; Youngson and Simpson 1981).

Thyroxine (T_4) increases swimming and aggression in both sexes of dwarf gourami (Forselius 1957). Chronic T_4 treatment in substrate-spawning cichlids seems to increase locomotor activity, aggression and territory size, while the thyroid-blocker thiourea has the opposite effect (Spiliotis 1974). In the blue acara, preliminary data confirm these results and furthermore indicate that the response to thiourea cannot be dismissed as due to toxicity since potassium perchlorate has a similar effect (Munro, in prep.). This may be an indirect product of changes in locomotor activity resulting in changes in encounter rate (Figure 9.4).

Melatonin

This hormone is secreted by the pineal gland (Figure 9.1) and the retina, especially during the dark phase of a normal light cycle. Possible roles include the control of circadian rhythms and reproduction (Munro and Dodd 1983).

In goldfish, melatonin has a sedative action (Satake 1979). Rasa (1969, 1981) briefly reports that this hormone controls territorial aggression in a damselfish, where defence of a feeding territory on the coral reef is necessary long before gonad development. The pineal is thought to secrete melatonin and mediate pituitary function (Fenwick 1970).

The Control of Parental Behaviour

This complex of action patterns includes brood defence, which may be thought of as a combination of resource- (territory) and self- (offspring) defence.

Of the mammalian adrenohypophysial hormones tested by Blum (1974), only prolactin induced some elements of parental behaviour in male sticklebacks and sunfish, and both sexes of substrate-spawning cichlids. Although eggs were absent, apparent egg-fanning behaviour was scored in these experiments on reproductively-naive fish. Noble, Kumpf and Billings (1938) found a similar effect of prolactin only in reproductively-experienced cichlids presented with eggs.

Blum's conclusion, that prolactin is the parental hormone, is debatable. Smith and Hoar (1967; Baggerman 1968) considered that testosterone, rather than prolactin, was the parental hormone in male sticklebacks. In cichlids, the 'parental fanning' induced by prolactin was not accompanied by any increase in aggression, another feature of brood care, indicating that if prolactin is involved in the control of parental behaviour, it is not the only factor.

Future studies of prolactin as a parental hormone require a much more rigorous methodology. For example, teleost prolactins should be used for injection: 'fanning behaviour' may be a toxic reaction to the mammalian hormone and other substances (Noble *et al.* 1938; Gona 1979), involving locomotor or respiratory problems. Also, increased prolactin secretion during the brooding phase has yet to be demonstrated. Metuzals, Ballintijn-de-Vries and Baerands (1968) claimed that the prolactin cells were stimulated during the parental phase, but the cells in question are probably growth hormone-secreting (Munro and Dodd 1983). An alternative source of biologically-active prolactin may be certain cells in the preoptic area which innervate the pituitary (Munro, in prep.).

Fiedler (1974) reported that male anabantoids built bubble-nests after injections of prolactin in combination with testosterone. Noble *et al.* (1938) found that corpus luteum extract, a source of progesterone, was almost as effective as mammalian prolactin in inducing fanning by cichlids of introduced eggs, and progesterone also increases aggression in isolated blue acara (Munro and Pitcher, unpubl. obs.). Therefore, whether or not prolactin is involved, progestagens may be parental hormones, in cichlids at least. In salmonids, plasma progestagens peak after ovulation (Chapter 11), and in sticklebacks have been implicated in the maintenance of post-ovulatory oocytes until spawning (Lam, Nagahama, Chan and Hoar 1978). Therefore, there is at least circumstantial evidence that progestagen levels may be high in the post-spawning female cichlid.

The Effects of Behaviour on the Endocrine System

The intimate reciprocal influences of the endocrine system with behaviour in mammals (Leshner 1978, 1981) are also found in fish. For example, ovulation in the brown acara occurs only as a consequence of courtship (Chapter 7). In mammals, victory in an agonistic encounter stimulates testosterone, whereas defeat stimulates corticosteroid production, with effects on subsequent agonism. Unfortunately, there is no direct evidence for this in fish: the technical difficulties of obtaining reliable measures of circulating hormones remain to be solved. However, there is some indirect evidence.

The most dominant individual in an all-female community of *Haplochromis* (Wapler-Leong and Reinboth 1975) and *Pseudotropheus* (Munro, unpubl. obs.) assumes some features of male colouration which are androgen-dependent (Reinboth 1972). Thus social dominance may stimulate testosterone secretion in cichlids, at least. A similar relationship, with or without changes in cortisol secretion, may explain the situation amongst protogynous hermaphrodite reef fish, where a male dominates a harem: removal of the male results in the most dominant female changing sex (Shapiro and Boulon, 1982). This basic explanation may be applicable, with modifications, for the situation in protandrous hermaphrodites.

Conclusions

We have strong evidence that agonistic behaviour is controlled by several endocrine axes in teleosts and there is some suggestion of a recriprocal influence of behaviour on hormone secretions. Unfortunately, but not surprisingly, only the rather specialised teleosts with highly evolved reproductive behaviour and brood care have been studied in detail in the laboratory and it is likely that more generally applicable conclusions might result from the study of agonism in more typical teleosts such as salmonids, cyprinids, gadoids, or the less specialised members of the perch order.

The pituitary—gonad axis drives agonism associated with reproduction in one of at least two fashions, depending on whether only males exhibit reproductive resource-defence (sticklebacks, sunfish) or both sexes carry out such behaviour (cichlids).

The pituitary—adrenocortical axis is also involved in the control of agonistic behaviour, but here it is likely that cortisol drives self-defence and reactions to environmental and social stress in a non-reproductive context.

Defence of offspring in teleosts, where it has evolved, does not appear to be driven directly by prolactin, but requires non-aromatised androgens and/or progestagens. Some non-defence aspects of brood care may be driven by prolactin in synergism with the steroid hormones.

Very little is known directly of the mechanism by which these hormones alter agonistic behaviours. Suggestive sites sensitive to some hormones have been demonstrated in a few teleost brains, but wider studies are required. Mammalian work has shown that hormones can affect the receipt and processing of sensory information, motivational decisions, and motor output, but in fish these studies are in their infancy. One major descending pathway involved in the control of agonistic behaviour has been demonstrated in teleosts and some neurons in it are known to be targets for sex steroids, but as yet direct experimental confirmation is lacking.

Acknowledgement

A.D. Munro would like to thank SERC for financial support.

10 OOCYTE DEVELOPMENT PATTERNS AND HORMONAL INVOLVEMENTS AMONG TELEOSTS

Victor deVlaming

Reproduction is obviously an essential link in the life history of fish, perpetuation of any species depending on successful reproduction. A consideration of reproductive physiology and the control mechanisms involved in reproductive processes is therefore imperative. The primary objective of this chapter is to present an introduction to the process of gamete formation and its control among female teleostean fishes. Limiting the discussion to female teleosts is still an enormous task consequent to the large number of species and incredible diversity in this group. Even though there are several viviparous teleosts, the majority of species are oviparous. Only oviparous species will be considered here.

Most teleostean species manifest an annual rhythm of breeding which, in many cases, is synchronised or controlled by environmental factors (deVlaming 1974; see also Chapter 11). Breeding at a specific time necessitates rather precise regulation of the reproductive system, especially in females where large numbers of yolked ova are generated. Given that eggs are the largest cells formed in the animal kingdom (e.g. ovulated eggs in the catfish, *Bagre marinus*, have an average diameter of 20 mm) and the incredible numbers produced by some species (a large ocean sunfish, *Mola mola*, can produce 28×10^6 eggs per year), it becomes clear that ovarian recrudescence requires mobilisation, synthesis, and transfer to the developing oocytes of immense quantities of materials. Mediation of the environmental effects, governing of the reproductive system, and mobilisation of reserves for reproduction are primarily under the auspices of the endocrine system. I shall, therefore, briefly summarise that which is known about hormonal control of the various aspects of oocyte growth, and draw attention to the aspects of oocyte development where information about hormonal regulation is lacking.

Wallace and Selman (1981) recently presented a thorough and lucid review of the cellular and dynamic aspects of oocyte growth among teleosts. Rather than duplicate their review, my aim is to elaborate on particular characteristics of oocyte development, and to present a review complementary to the Wallace and Selman work. The mechanisms of oocyte growth appear similar for most teleostean species, the greatest diversity occurring in recruitment and timing events. Therefore this chapter will devote particular attention to those topics.

Primary Growth of Oocytes

Oogonia are small, rounded cells with a high nucleus to cytoplasm ratio. Generally these cells occur in nests, but sometimes singly in the stroma of the ovigerous folds. Oogonia proliferate by mitoses; this proliferation frequently shows a seasonal peak in species with an annual spawning cycle. In species with protracted or continuous breeding seasons oogonial divisions may occur repeatedly or continuously throughout the year. There are conflicting reports as to a possible endocrine role in regulating oogonial mitoses. Some researchers maintain that pituitary gonadotrophin, acting directly or indirectly (via steroids), promotes oogonial proliferation, whereas other investigators do not support this contention (deVlaming 1974).

The transformation of oogonia into primary oocytes (oogenesis) has not been thoroughly investigated among teleosts, yet the present consensus is that chromosomes become arrested at the prophase (diplotene phase) of the first meiotic division (Tokartz 1978). Soon after this transformation squamous follicle cells begin to encapsulate the oocyte, which consists of a centrally located nucleus surrounded by a thin cytoplasmic layer. During the primary growth phase the nuclear-to-cytoplasm ratio decreases dramatically as the oocyte volume increases a few hundred-fold. Nonetheless, the absolute size of the oocyte nucleus increases, and multiple nucleoli appear (perinucleolar stage). This proliferation of nucleoli presumably reflects ribosomal gene amplification (e.g. Vlad 1976), although recent studies are not supportive of this concept. In the few species examined, lampbrush chromosomes are formed during the primary growth phase, and they disappear prior to the nuclear (i.e. germinal vesicle) breakdown of oocyte maturation (e.g. Baumeister 1973).

During the primary growth phase basophilic material (containing ribonucleoproteins), presumably extruded through the nuclear envelope pores, accumulates in the perinuclear cytoplasm (Riehl

1978; Guraya 1979). Organelles come to enclose the basophilic aggregates, and these complexes have been termed Balbiani bodies or 'yolk nuclei'. Prior to the appearance of cortical alveoli in primary oocytes the Balbiani body is translocated to the periphery of the cytoplasm and disperses. Although the role of the Balbiani body is yet to be completely defined, it may act as a centre for formation of organelles (Toury, Clérot and André 1977; Guraya 1979). The prevalent hypothesis is that the primary growth phase can occur independently of pituitary hormone control. Little is known concerning the possible role of steroid hormones in this phase of oocyte growth. Studies with *Leptocottus armatus* reveal that plasma oestradiol-17β levels are at a seasonal low concomitant with the oocyte primary growth phase (deVlaming, unpubl. obs.).

The follicular wall in growing oocytes consists of a chorion, granulosa layer, basement membrane (acellular), and theca (Guraya 1978). During the early phases of oocyte growth a chorion is not present and the granulosa cells are adjacent to the oocyte. As the oocyte grows the granulosa cells increase in number to form a continuous single layer around the oocyte. Contradictory reports exist concerning the potential role of granulosa cells in steroid hormone biosynthesis. Ultrastructural and/or histochemical studies with some species suggest that the granulosa cells have steroidogenic capabilities (Sundararaj 1981). These contradictory data may be the result of changes in steroidogenic activity of the granulosa cells with the stage of oocyte development.

A chorion (vitelline membrane, zona radiata, zona pellucida) forms between the granulosa cells and oocyte during the secondary growth phase of the follicle. The origin of the chorionic protein and carbohydrate components is controversial, its synthesis being attributed to the oocyte, the granulosa cells, or both (Laale 1980). Most recent research reveals that the chorion is a multilamellar structure with a dual origin, the 'primary envelope' being formed by the oocyte and the 'secondary envelope' a product of the granulosa cells. These two 'envelopes' come to fuse in the course of oocyte development. Granulosa cells develop cytoplasmic processes which penetrate the chorion to variable degrees, depending on the stage of oocyte growth. Microvilli, which cover the surface, are generated by the oocyte. The follicular cell processes and the oocyte microvilli traverse pores in all regions of the chorion. These projections allow communication between the granulosa cells and oocytes (Wallace and Selman 1981).

Outside a basement membrane, the theca surrounds the follicular epithelium. The thecal layer consists of fibroblast-like cells, and is infiltrated by a capillary network. In several teleost species there is evidence that specialised thecal cells are steroidogenic, whereas other investigators deny an endocrine role for these cells (Sundararaj 1981). There may be species differences with regards to steroidogenic capacity of thecal cells; before rejecting a steroid biosynthetic role of the theca in any species, it is essential to examine these cells in all phases of oocyte growth. Steroidogenic capabilities in granulosa cells and thecca are not mutually exclusive, even though synthetic activity may not be simultaneous. Among mammals thecal cells produce androgens which are aromatised to oestrogens by granulosa cells. The possibility of a homologous thecal—granulosa interaction among teleosts is worthy of scrutiny.

Secondary Growth of Oocytes

Cortical Alveoli Formation

The secondary growth phase of oocytes is gonadotrophin-dependent, as shown in hypophysectomy-replacement therapy experiments (deVlaming 1974). The so-called 'yolk' vesicles (PAS-positive) are the first structures to appear in the oocyte cytoplasm during this growth phase. The mucopolysaccharide or glycoprotein contents of the 'yolk' vesicles are synthesised within the oocyte (autosynthetic). Initially the vesicles appear in the outer half of the cytoplasm, enlarging as they move to take a position subjacent to the oolemma. The endoplasmic reticulum and Golgi elements seem to be involved in elaboration of 'yolk' vesicles, and in some teleost species two populations of these structures can be detected (Iwamatsu and Keino 1978). Most probably 'yolk' vesicles give rise to cortical alvcoli. In the later stages of the secondary growth phase cortical alveoli lie subjacent to the oocyte membrane. The cortical alveoli fuse with the oolemma, releasing their contents (cortical reaction) into the perivitelline space at the time of fertilisation. Assuming that the 'yolk' vesicles are the cortical alveoli precursors, the contents of these structures cannot be termed yolk in a true sense (see also Chapter 11).

Oil droplets begin to accumulate in the cytoplasm of oocytes in some teleosts (particularly marine species) about the same time as the cortical alveoli precursors. These smaller oil droplets generally coalesce into several larger droplets towards the end of vitellogenesis or during oocyte maturation.

Vitellogenesis

The accumulation of yolk is responsible for the enormous enlarge-ment of oocytes during the secondary growth phase. Crystalline yolk has been described in the oocyte granules of some teleost species, but in most fish yolk proteins accumulate in fluid-filled yolk spheres (Wallace 1978; Wallace and Selman 1981). In some species these yolk spheres maintain their integrity throughout oocyte develop-ment, whereas in other teleosts the spheres fuse at some stage to yield a continuous mass of fluid yolk. The transparency of eggs in certain species is due to this fusion of yolk spheres. Fusion of yolk spheres can occur soon after they begin to appear in the ooplasm, during the later stages of vitellogenesis, or during oocyte maturation. Yolk in piscine oocytes consists of phosphoproteins and lipoproteins (Wallace 1978; deVlaming, Wiley, Delahunty and Wallace 1980). More specifically, teleostean yolk proteins consist of a lipovitellin fraction and phosvitin fraction. Although not thoroughly validated it seems that multiple molecular weight forms of both phosvitin and lipovitellin exist in the oocytes of teleosts (deVlaming *et al.* 1980).

In a number of piscine species, female-specific serum lipoproteins or proteins have been detected during the vitellogenic phase of the annual reproductive cycle (deVlaming *et al.* 1980; Wallace and Selman 1981). In several species antisera prepared against oocyte or egg extracts cross-react with serum of vitellogenic females, but not with serum of males or immature females. Such observations lead one to conclude that yolk proteins are of an extra-ovarian origin. Furthermore, oestrogen administration induces the appearance of these female-specific proteins in the serum of male and immature female fish. For example, chromatographic profiles of plasma collected from female goldfish in the process of yolk accumulation, from normal males, and from oestrogen-treated males are presented in Figure 10.1. These data illustrate that a phosphoprotein is present in the serum of vitellogenic females which is absent in males, but its presence in males can be induced by oestradiol-17β treatment. Thus, among teleosts, as in other oviparous vertebrates, a female-specific serum phospholipoglycoprotein (vitellogenin) has been identified as the egg yolk precursor.

Vitellogenin is synthesised by the liver in response to oestrogen stimulation, released into the circulation, and specifically seques-tered, under gonadotrophin control, by developing oocytes. Pituitary gonadotrophin enhances vitellogenin transfer from blood into oocytes by eliciting micropinocytotic activity at the oocyte surface

Figure 10.1: Chromatography on DEAE-cellulose of plasma from goldfish, *Carassius auratus*, previously injected with [^{32}P]orthophosphate; (a) 0.5 ml plasma from vitellogenic females; (b) 0.6 ml plasma from normal males; and (c) 0.6 ml plasma from oestrogen-injected males. The absorbance trace (solid lines) indicates a component at elution position 0.68 which is present in females and oestrogen-treated males, but is absent in males. This component (vitellogenin) is specifically labelled with [^{32}P]ortho-phosphate, as indicated by labelling (●—●—●) found in TCA-precipitated, de-lipidated samples of the eluant.

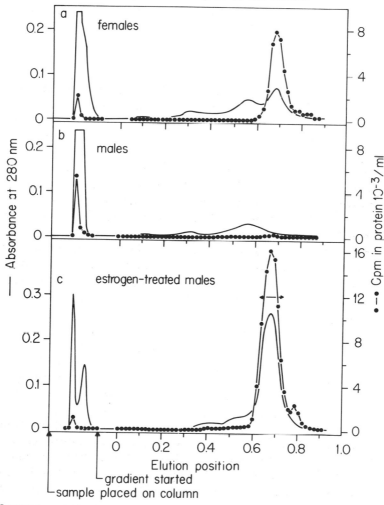

Source: from deVlaming *et al.* 1980

(Wallace and Selman 1981). Vitellogenesis is gonadotrophin-dependent not only because oocyte uptake of the yolk precursor is dependent on a pituitary factor, but also because ovarian oestrogen synthesis is evoked by gonadotrophin (Sundararaj 1981).

Although limited data are available, multiple vitellogenin polypeptides are apparently processed into multiple yolk protein polypeptides within growing oocytes of teleostean fishes. Piscine vitellogenins may have lower molecular weights, higher lipid content, and lower phosphorus content than the yolk precursors of other oviparous vertebrates (deVlaming *et al.* 1980). More extensive studies are needed to corroborate these hypotheses.

Oocyte Maturation and Ovulation

When oocytes reach a critical size, which is species-specific, the sequestering of yolk precursor is curtailed. Although not clearly demonstrated, it appears that in some species oocytes can be 'arrested' for a time at the post-vitellogenic stage before recruitment into maturation and ovulation; in other species oocytes may move more directly into maturation. Whether or not there is an arrestment in the post-vitellogenic stage, the prevailing theory is that recruitment into maturation requires appropriate hormonal stimulation (Kanatani and Nagahama 1980).

Oocyte maturation consists of multiple biochemical and physical changes, but the primary event is the resumption of meiosis. Without much documentation it is commonly alleged that the chromosomal activity proceeds to the metaphase of the second meiotic division. Resumption of meiosis is heralded by a peripheral migration of the germinal vesicle (nucleus) and by dissolution of the germinal vesicle membrane. Germinal vesicle breakdown (GVBD) is a commonly used indicator of oocyte maturation. In some species coalescence of yolk granules (oocytes become more transparent) and/or of lipid droplets accompanies the nuclear events. Concomitant with maturation in many teleosts, especially marine forms with pelagic eggs, is oocyte enlargement due to hydration. Hydration can, in some species, result in a 300-400 per cent increase in oocyte volume. The mechanisms of this hydration are not understood, but the phenomenon seems to be hormonally triggered (Wallace and Selman 1981).

The contention at the present time is that oocyte maturation and ovulation are dependent on pituitary gonadotrophin. Ovarian pieces have been tested for responsiveness to gonadotrophins *in vitro*; if the follicles are of a critical size, various mammalian and piscine gonadotrophins evoke, in most species tested, GVBD. In some teleostean species a dramatic increase of plasma gonadotrophin levels ('ovulatory surge') occurs just prior to oocyte maturation and ovulation (Stacey *et al.* 1979). The prevailing hypothesis asserts that this increase in plasma gonadotrophin elicits production of a maturational steroid hormone which initiates oocyte maturation (Kanatani and Nagahama 1980).

Interspecific variations exist in certain aspects of the endocrine control of oocyte maturation. The site of maturational hormone (MH) synthesis may differ; follicular cells are probably responsible for MH production in the majority of teleosts. The inter-renal gland may produce a MH in a few species. The inability of a gonadotrophin to promote *in vitro* maturation of post-vitellogenic oocytes can indicate an extra-ovarian source of the MH. Such data would not be conclusive, however, since hormones which have a 'permissive' function or which synergise with gonadotrophin (or a MH of ovarian origin) could be absent *in vitro*. The steroids which induce oocyte maturation are not necessarily the same from species to species.

Maturational hormones are general C_{21} steroids, and can be placed into one of two broad categories — corticosteroids (especially 11-deoxycorticosteroids) or progesterone-related steroids (Jalabert 1976; Kanatani and Nagahama 1980). Common between the progestogens and deoxycorticoids is the absence of an oxygen function on C11 of the steroid ring structure. Progesterone or related steroids, especially 17α-hydroxy-20β-dihydroprogesterone and 17α-hydroxyprogesterone, are the most potent (*in vitro*) promoters of GVBD in *Oncorhynchus kisutch, Salvelinus fontinalis, Salmo gairdneri, Esox lucius, Carassius auratus*, and *Cyprinus carpio* (Kanatani and Nagahama 1980). 17α-hydroxy-20β-dihydroprogesterone has been identified and its concentration determined in the blood of a few species, and levels increase dramatically near the onset of oocyte maturation (e.g. salmonids, Chapter 11). In *Heteropneustes fossilis* 11-deoxycorticoids are highly effective maturational hormones (Sundararaj 1979). Both progesterone-related steroids and 11-deoxycorticosterone (DOC) are potent (*in vitro*) as MHs in *Perca flavescens* (Goetz and Theofan 1979) and *Brachydanio rerio* (VanRee, Lok and Bosman 1977). In some species where DOC has been shown to be an active MH, progesterone derivatives have not been tested. A MH should be able to induce maturation in post-vitellogenic oocytes stripped of the follicular layers; more studies oriented toward such experiments could prove informative. It is worth remembering that the steroid most effective at inducing GVBD *in vitro* is not necessarily the *in vivo* MH.

Sundararaj (1979) reported that gonadotrophin stimulates both the ovarian production of epipregnanolone, a progesterone metabolite, and inter-renal synthesis of cortisol in *Heteropneustes*. He hypothesises that epipregnanolone sensitises post-vitellogenic oocytes to the presumed MH (cortisol). This claim could be strengthened by testing cortisol and epipregnanolone on denuded, post-vitellogenic oocytes. Almost certainly corticosteroids play some role in oocyte maturation and ovulation among teleosts. Indeed, in some species, but not others, there is a 'surge' of cortisol in the plasma a few hours prior to ovulation (Cook, Stacey and Peter 1980). That cortisol sensitises follicles to gonadotrophin-stimulated synthesis of MH seems probable. Another likely possibility is that cortisol sensitises the oocyte to MH or synergises with MH. The site of MH synthesis, ovarian or otherwise, remains equivocal and the molecular events evoked by MHs await elucidation.

Concomitant with maturation or after maturation is complete, oocytes are ovulated into the ovarian lumen. The mechanics of ovulation among teleosts deserve considerably more attention. Whether oocyte hydration plays a role in ovulation is debated, but of probable significance is the onset of follicle cell dissociation just prior to ovulation. Microfilaments located in follicular thecal cells are accredited a cell-movement function during ovulation by some. Because prostaglandins stimulate ovulation in various species (see below), some researchers conclude that the process is accomplished by smooth muscle fibres in the theca.

Ovulation does not seem to require a separate endocrine signal in some teleosts (i.e. it follows irrevocably once maturation has been initiated). Moreover, maturation and ovulation can be dissociated in some species (e.g. *S. fontinalis*), but not in others (*H. fossilis* and *P. flavescens* — Sundararaj 1981). Frequently steroids (which alone will induce GVBD) will not promote ovulation *in vitro*. Gonadotrophin, therefore, may be directly or indirectly necessary for ovulation in some teleosts. Jalabert (1976) surmised that the MH stimulates the synthesis of ovulatory agents, prostaglandins, by follicular cells; prostaglandins E_2 and $F_{2\alpha}$ have proved to be potent inducers of ovulation in several species. One interesting hypothesis alleges that prostaglandins produced in the ovary at the time of ovulation enter the circulation to act at a CNS site to induce spawning behaviour (Stacey 1981). If the MH elicits prostaglandins synthesis, however, one would expect that ovulation would follow steroid-induced maturation *in vitro*; as mentioned above, such is not always the case. Much emphasis has been placed on the prostaglandins as ovulatory agents (Stacey and Goetz 1982). There is experimental evidence that catecholamines, acting via α-adrenergic receptors, can induce ovulation in some species (Jalabert 1976).

Spawning

It should be emphasised that ovulation and spawning are separate events under different control mechanisms. Many teleost species go through multiple spawning acts per breeding season; whether all eggs oviposited at each spawning act are ovulated at approximately the same time or whether the oviposited eggs of a single spawning act are the product of multiple ovulations (over a period of time) is not always clear. Moreover, fractions of an ovarian stock of ovulated eggs may be intermittently spawned, so separate ovulatory events may result in accumulation of a stock of eggs which could conceivably be spawned *in toto* or in fractions.

Is there a relationship between the frequency of spawning or the duration of the spawning season and the type of oocyte development? Species with both group-synchronous and asynchronous oocyte development (see below) are known to have multiple spawnings per female in a breeding season and protracted spawning seasons. It should be clearly understood that a protracted spawning season cannot always be equated with multiple spawns for each female. Protracted spawning seasons could simply reflect a lack of population synchrony in terms of gonadal development. Extreme caution must be applied when attempting to predict the number of spawns per female in a spawning season based on the number of oocyte clutches detectable in the ovary. To determine the actual dynamics of spawning it is essential to observe individual females.

The terminology employed to describe spawning activity can be misleading. For example, fractional spawner and multiple spawner are frequently used synonymously. The term multiple spawner is generally applied to a species in which a female spawns more than once in a spawning season. The term fractional spawner has been used to refer to species which spawn a part of an ovulated clutch or which mature, ovulate, and spawn a part of a post-vitellogenic clutch at intervals over a relatively short period. There is little conclusive evidence, however, for partial recruitment of a post-vitellogenic clutch or of a clutch in the late stages of vitellogenesis.

Follicular Atresia and Post-ovulatory Follicles

Relatively small numbers of vitellogenic oocytes which fail to be ovulated prior to a spawning, or an entire clutch of oocytes undergoing vitellogenesis when environmental conditions become unfavourable, can become atretic (i.e. degenerate). During the normal process of ovarian recrudescence a relatively small percentage of oocytes usually become atretic; oocyte atresia is uncommon in healthy (non-stressed), well-fed fish. In the process of atresia, follicular epithelial cells, and possibly thecal cells, invade the oocyte to reabsorb its contents. In some species follicular cells proliferate and hypertrophy to form a compact, well-vascularised structure (corpus atreticum — referred to on occasion, and perhaps misleadingly, as a pre-ovulatory corpus luteum). Contrasting conclusions, which could be based on interspecific differences, have been reached regarding the steroid hormone biosynthetic capacity of these corpora atretica. Some researchers concluded that the corpora atretica are solely involved in removing degenerative oocytes, and that they do not resemble mammalian corpora lutea, whereas others contend that corpora atretica possess an endocrine function (see discussion in Kagawa, Takano and Nagahama 1981).

In most teleostean species the post-ovulatory follicles are relatively short-lived. Corpora lutea-like structures are formed in post-ovulatory follicles of some species, but not in other species. Limited histochemical and correlative data imply that post-ovulatory follicles have an endocrine role (i.e. steroid synthesis — Kagawa *et al.* 1981). More extensive and thorough investigations are obviously needed if we are to understand the potential functions of corpora atretica and post-ovulatory follicles among teleosts.

Dynamics of Oocyte Development

The foregoing describes generally the processes of oocyte growth; these processes are fairly similar among most oviparous teleost fish.

Nonetheless, the basic processes of oocyte development are modified in a variety of ways among species to assure successful reproductive strategies. For example, there is considerable interspecific variation in the size of eggs produced (see Wallace and Selman 1981). Another aspect of ovarian development is the time frame of oocyte growth and of the various phases therein. The interval between spawning seasons is frequently annual, yet within this annual scheme there are species which spawn once per year and others which spawn several times during the breeding season. There are records of species which spawn every other year and of species which, during the breeding season, spawn every few days. Yet to be elucidated are the similarities and differences in the endocrine control of reproductive processes among the species with such diverse reproductive strategies.

Patterns of oocyte development can be detected by plotting the frequency of oocytes in an ovarian sample as a function of oocyte diameter. Analyses of such plots have revealed three basic patterns of oocyte development among teleosts. Marza (1938) appears to have been the first to describe the *synchronous, group-synchronous,* and *asynchronous* patterns of oocyte development. Frequently investigators pool data from several ovaries rather than analysing the frequency distribution of an individual female. This pooling can obscure the actual pattern of oocyte development, especially if there is not total synchrony of females within the population sampled. Another common means of reporting data is to plot mean oocyte diameter in a sample as a function of month (or some other expression of time). Such reports infrequently include a reference as to whether this mean is of one clutch (synchronous group with relatively uniform diameters) of oocytes or is of all oocytes in the ovary. Attention should be called to the fact that oocytes in different parts of the ovary may differ in stage of development and thus in size.

Synchronous Oocyte Development

In ovaries where development is synchronous all oocytes, once formed, grow in unison; recruitment of a second clutch does not occur. Such ovaries would be characteristic of those teleosts, such as anadromous salmon and catadromous eels, which die after spawning. Synchronous oocyte development may be more common among or restricted to the teleost super-orders which are considered to have had an early evolutionary origin (e.g. Elophomorpha and Protacanthopterygii — Table 10.1). The reader should be aware that not all species in teleost groups which are presumed to have an early evolutionary origin have synchronous oocyte development (Table 10.1). Whether or not endocrine control of ovarian development in such species is similar to teleosts with asynchronous or group-synchronous oocyte development is a provocative question not yet addressed.

Group-synchronous Oocyte Development

Where there is group-synchronous oocyte development at least two populations (clutches) of oocytes can be distinguished in the ovary of an individual at some time during the reproductive cycle. This is undoubtedly the most common ovarian type among teleost fishes. Group-synchronous oocyte development varies in terms of the number of clutches which can be distinguished in a single ovary and the 'distinctness' (degree of synchrony or uniformity of diameter) of the clutch. So there are species variations in the number and frequency of recruitments into vitellogenesis from the 'yolk' vesicle stage or the primary growth stage. At one extreme (e.g. *Perca fluviatilis* and many salmonids) there is recruitment out of the primary growth stage into vitellogenesis of one clutch per year (i.e. spawning season). On the other hand, four clutches of oocytes can be discerned at a single time in an individual ovary of the medaka (*Oryzias latipes*). Generally the leading clutch (population with largest diameters) oocytes are much more uniform (synchronous) in diameter than the smaller, more heterogenous previtellogenic oocytes (Figure 10.2). The uniformity of oocyte diameters in the leading clutch also is species variable. In some species (e.g. *Gasterosteus* — Figure 10.2; *Leptocottus armatus* — see below) the leading clutch is relatively synchronous, whereas in other species (e.g. haddocks and herrings) the leading clutch is much less distinct. The existence of non-uniform clutches in the ovaries can lead to variable nomenclature as to type of oocyte development. For example, that goldfish (*Carassius auratus*) manifest asynchronous oocyte development has been contended. Another interpretation is that oocytes of the goldfish ovary develop in clutches which are not highly uniform in diameter.

One of the most common patterns in group-synchronous ovaries is the appearance of three clutches of oocytes just prior to and during the spawning season (Figure 10.3). The leading clutch is post-vitellogenic or in the process of vitellogenesis, a second clutch is 'arrested' in the cortical alveoli ('yolk' vesicle) stage, and there is a third clutch of non-yolky oocytes in the primary growth phase. In *Gasterosteus* and *Leptocottus*, as in many other species, recruitment

Table 10.1: Teleost taxonomic groups in which one or more species has been reported to have synchronous, group-synchronous, or asynchronous oocyte development.

Synchrony	Group-synchrony	Asynchrony

(1) Elopomorpha
    Anguilliformes
        Anguillidae
	(1) Clupeomorpha
	    Clupeiformes
	        Clupeidae
	        Engraulidae
		(1) Clupeomorpha
		    Clupeiformes
		        Clupeidae

(2) Protacanthopterygii
    Salmoniformes
        Salmonidae
	(2) Protacanthopterygii
	    Salmoniformes
	        Salmonidae
	        Esocidae
	        Osmeridae
		(2) Ostariophysi
		    Cypriniformes
		        Cyprinidae
		    Siluriformes
		        Sisoridae

	(3) Ostariophysi
	    Cypriniformes
	        Cyprinidae
	        Catostomidae
	    Siluriformes
	        Bagridae
	        Claridae
	        Heteropneustidae
	        Siluridae
		(3) Paracanthopterygii
		    Gadiformes
		        Gadidae
		
		(4) Atherinomorpha
		    Aterinoformes
		        Cyprinodontidae

	(4) Paracanthopterygii
	    Gadiformes
	        Gadidae
		(5) Acanthopterygii
		    Gasterosteiformes
		        Syngnathidae
		    Scorpaeniformes
		        Cyclopteridae
		    Perciformes
		        Blenniidae
		        Scombridae

	(5) Atherinomorpha
	    Aterinoformes
	        Cyprinodontidae
	        Oryziatidae	

	(6) Acanthopterygii
	    Gasterosteiformes
	        Gasterosteidae
	    Scorpaeniformes
	        Cottidae
	        Cyclopteridae
	        Scorpaenidae
	    Periformes
	        Serranidae
	        Mugilidae
	        Percidae
	        Istiphoridae
	        Centrarchidae
	        Cichlidae
	        Gobiidae
	        Sparidae
	        Scombridae
	        Sciaenidae
	        Nototheniidae
	    Pleuronectiformes
	        Pleuronectidae
	        Bothidae	

of the second clutch from the 'yolk' vesicle stage into vitellogenesis does not normally occur until maturation is initiated in the leading clutch (Figures 10.2, 10.3). As indicated above, vitellogenesis is a gonadotrophin-dependent process. An intriguing endocrine question concerns the apparent refractoriness to tonic gonado-trophin levels of the second clutch, as compared to the leading clutch, of oocytes. In some teleosts oocyte maturation and ovulation are elicited (indirectly) by a dramatic increase in plasma gonadotrophin levels (ovulatory surge — Stacey *et al.* 1979). Possibly the ovulatory surge of gonadotrophin is also essential for recruiting a second clutch of oocytes into vitellogenesis. This hypothesis is supported by the experiments of Wallace and Selman (1979). These investigators observed that treatment of *Apeltes quadracus* with a gonadotrophin (human chorionic gonadotrophin) recruited follicles into vitello-genesis regardless of the stage of oocytes in the leading clutch. As the above indicates, recruitment of oocytes into vitellogenesis can be from the 'yolk' vesicle stage. In other species with group-synchronous ovarian development recruitment can be directly from oogonia or from a population of oocytes 'arrested' at the end of the primary growth phase. Whether the endocrine mechanisms involved in all these forms of recruitment are similar is yet to be resolved.

Ovarian development in the staghorn sculpin, *Leptocottus armatus*, is characterised by group-synchronous oocyte development (deVlaming, unpubl. obs.). For five months of the year the ovaries of some or all fish in the population studied (and for three months of all of the fish sampled) contained three distinct clutches of oocytes (Figure 10.3). Ovaries of some fish sampled during March were characterised by follicular atresia. By April this ovarian regression was more extensive. Only non-yolky oocytes were noted in ovaries of fish collected from May through August. Recruitment of an oocyte clutch into the cortical alveoli stage was evident during September and October; this clutch was then recruited into vitellogenesis during November. It is clear from the data presented in Figure 10.3 that oocyte development in the population of *Leptocottus* studied is not synchronous (i.e. ovaries from individuals are not in the same stage of development). Similar results have been recorded for sticklebacks (Wallace and Selman 1979). The data presented in Figure 10.2 show frequency distributions of oocytes in ovaries from *Gasterosteus aculeatus* and *Apeltes quadracus* sampled at the same time. The diameter of oocytes in the vitellogenic clutch (i.e. leading clutch) of the different fish is variable (individuals in the population are not synchronous), indicating that recruitment of clutches is not likely to be initiated by a physical environmental event (Figure 10.2)

Group-synchronous oocyte development has been observed in representative species of many teleostean taxonomic groups (Table 10.1). This table is by no means a complete listing of all taxonomic groups examined nor is it my intention to imply that all species in these families have group-synchronous oocyte development. Of note is that representative species of all teleost super-orders except Elopomorpha and Osteoglossomorpha, which theoretically represent an evolutionary line with an early origin, have been studied. Valuable insights into ovarian development patterns and their control mechanisms could result from investigations with species representing these two super-orders, as well as with more paracanthopterygiian species.

Asynchronous Oocyte Development

A frequently made but erroneous assumption is that species with protracted spawning seasons and multiple spawns per female have asynchronous oocyte development. *Fundulus heteroclitus* and *Oryzias latipes* are both multiple spawners with protracted spawning seasons; oocyte development in some populations of the former is asynchronous, but group-synchrony characterises the *Oryzias* ovary (Wallace and Selman 1981). Furthermore, there are populations of *Sarotherodon* (*Tilapia*) which spawn throughout the year, yet have pronounced oocyte clutches. In many cases where three oocyte clutches have been identified in an individual ovary, multiple spawnings have been hypothesised, but not verified. Three recognisable oocyte clutches in the ovary of a recrudescing fish is not necessarily indicative of 2-3 spawns during that breeding season; more than three spawns with several recruitments have been reported (Wallace and Selman 1979) and a single spawn with reabsorption of a clutch has been documented (Mackay and Mann 1969)

Oocytes of various stages, without pronounced clutches, are present in species with the asynchronous type of ovarian development. This type of oocyte development is characteristic of *Blennius pholis* (Figure 10.4), which is presumed to undergo multiple spawnings over a protracted (5-6 month) spawning season.

A random mixture of oocytes with a wide range of diameters can be identified in the ovary during the recrudescence and spawning phases (Figure 10.4). Most species with asynchronous oocyte development have protracted spawning seasons with multiple

Figure 10.2: Size-frequency distribution of ovarian follicles in 10 simultaneously field-collected *Gasterosteus aculeatus* (a) and *Apeltes quadracus* (b). Profiles are presented (i.e. ranked), solely for the sake of clarity, according to the largest follicles found in the ovary. Fish with ovulated eggs in the ovarian cavity are indicated. The dotted vertical lines separate primordial follicles from those undergoing vitellogenesis, while the dashed vertical lines separate the latter group from those undergoing maturation.

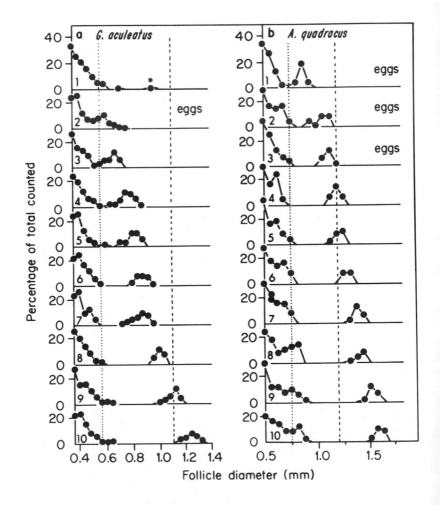

Source: From Wallace and Selman 1979. Reproduced by permission of the authors and Academic Press

Figure 10.3: Seasonal changes in ovarian development of the staghorn sculpin, *Leptocottus armatus,* from San Pablo Bay, California. Sample size in each month is 14-16. (a) Diameters of oocytes in recognisable clutches (O-clutch with smallest diameters; ●-intermediate clutch; and ▲-clutch with largest diameters). Arrow indicates point where recruitment occurs. (b) Seasonal variation in the clutch composition of ovaries. YO-clutch of oocytes showing yolk accumulation; CAO-clutch of oocytes in cortical alveoli stage; NYO-clutch of non-yolky oocytes without cortical alveoli; At-atretic follicles predominate in the ovary.

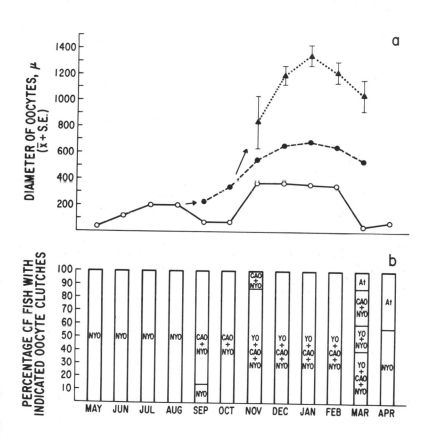

Figure 10.4: Size-frequency distribution of *Blennius pholis* ovarian follicles for January to March 1974 showing the asynchronous type of oocyte development. Hatched areas refer to distribution for fish collected in 1975. The data presented are pooled from several ovaries, but oocyte size distribution is similar in the ovaries of an individual fish. The numbers above histograms represent relatively distinct groups (clutches?) of oocytes; 1974 samples without parentheses and 1975 samples within parentheses.

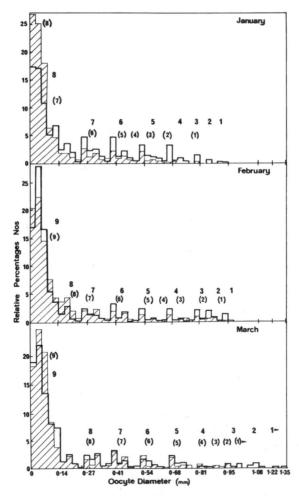

Source: from Shackley and King 1977. Reproduced by permission of the authors and Elsevier/North-Holland Biomedical Press

spawns per female. Distinguishing between ovaries with multiple clutch group-synchrony and those with asynchronous development could become a semantic exercise (see also the discussion of non-uniform group-synchronous clutches above).

Asynchronous oocyte development has not been recorded in as many taxonomic groups as group-synchrony (Table 10.1). That this type of oocyte development appears less common could be due to a paucity of studies, yet this seems unlikely. Asynchronous oocyte development presumably occurs among five of the eight teleostean super-orders. Asynchrony may not be as widespread taxonomically as indicated in Table 10.1. Ascription of asynchronous oocyte development to several species has been based on histological sections and this technique is *invalid* for assessment of development pattern.

In ovaries with asynchronous development there seems to be continuous recruitment (during recrudescence and the spawning season) into vitellogenesis and periodic recruitment into maturation/ ovulation. *Continuous* recruitment into vitellogenesis, such that many different-sized oocytes are accumulating yolk, could be the major distinguishing characteristic of asynchronous oocyte development. Such a situation presents intriguing endocrine questions. Experiments designed to define whether periodic gonadotrophin surges are needed to recruit vitellogenic oocytes into maturation and ovulation would be especially informative.

A particular type of oocyte development pattern is not necessarily common to all species in a given family. For example, ovarian organisation in some salmonids is group-synchronous, yet in other species it is synchronous. Asynchronous and group-synchronous oocyte development are recorded in various scombrids (tuna and mackerel). Members of the Cyclopteridae (Liparidae) are of interest because there are representative species distributed from the intertidal to abyssal regions. Many of the representatives in this family have protracted spawning seasons (and presumed multiple spawnings by individual females), continual breeding being alleged for some species. Stein (1980) presents data which illustrate group-synchronous to 'semi-asynchronous' types of ovarian organisation among cyclopterid species.

Summary

Taking a basic mode of oocyte development, multiple reproductive strategies have been elaborated by teleost fishes to match the diverse environments into which this group has radiated. Perhaps the greatest variations on the basic theme are in terms of timing of the various phases of oocyte development and of recruitment phenomena. Questions of how endocrine mechanisms compare between species with different types of oocyte development and with differing recruitment patterns are ripe for investigation.

Many aspects of female reproductive endocrinology are not thoroughly understood and thus could be profitably investigated. For example, potential hormonal involvement in oogonial proliferation and in the oocyte primary growth phase require attention. Conclusive demonstrations that 'yolk vesicles' are actually cortical alveoli, of their biochemical nature and of their sites and mechanisms of synthesis, are needed. The nature of endocrine regulation of cortical alveoli formation has not been elucidated. Whether two pituitary gonadotrophins exist in some teleostean fishes is still debatable; isolation of ovarian gonadotrophin receptors could be useful in solving this and other questions. That gonadotrophin promotes vitellogenin uptake by oocytes is yet to be clearly demonstrated. Direct examination of possible endocrine roles of postovulatory follicles, corpora atretica, the theca, and the granulosa could help untangle a confusing literature.

Several questions remain concerning endocrine mechanisms in oocyte maturation and ovulation. With the follicles of some teleosts, oocyte maturation can be evoked *in vitro* with steroid hormones or a gonadotrophin, but ovulation does not follow. This could be due to a lack of support (permissive) hormones. It seems essential that *in vitro* studies of oocyte maturation and ovulation include synergising or support hormones. For example, corticoids (especially cortisol) have been shown to potentiate the maturation-promoting effects of other hormones *in vitro*. The role corticoids play (and how) deserves investigation. More thorough investigation may reveal that corticoids are 'supportive' hormones rather than the actual oocyte maturational hormones. Indeed, it seems that prior to accrediting any hormone (or group of hormones) a MH status, an action on denuded oocytes must be shown. How gonadotrophin participates in oocyte maturation and ovulation has not been satisfactorily illustrated. Does gonadotrophin, for example, play a direct role in the synthesis of a MH, or the synthesis of an ovulatory hormone? Where is the MH synthesised and are the cytosolic receptors for it? A potential regulatory relationship between a MH and an ovulatory agent (e.g. prostaglandins) is yet to be illuminated. Because the ovarian wall in many species contains smooth muscle the potential roles of prostaglandins and neurohypophysial octapeptides (e.g. isotocin) in initiating oviposition beg for investigation.

Elusive still is the endocrine regulation of ovarian processes in species with asynchronous oocyte development. Ovulations of a few (even a single?) oocytes appear to occur frequently, and maybe rather continuously in *F. heteroclitus* during the spawning season (R. Wallace, pers. comm.). That a gonadotrophin surge initiates each ovulation seems unlikely. How maturation and ovulation are controlled in these species with asynchrony, as well as whether gonadotrophin is necessary to recruit oocytes into vitellogenesis, are provocative questions. The possibility that a diurnal surge of gonadotrophin evokes maturation/ovulation and recruitment into vitellogenesis of a few oocytes on a daily basis is worthy of investigation. With asynchronous oocyte development, is maturation/ovulation spontaneous once oocytes grow to a certain size, or is there a change in the competency of oocytes to respond to a maturation-inducing hormone?

Particularly informative would be detailed endocrine studies and comparisons on 'model species' representing the three basic types of oocyte development. Perhaps too little attention has been focused on the type of oocyte development when endocrine mechanisms are considered. In all cases, moreover, the dynamics of oocyte development must be examined and correlated with endocrine data. The sockeye salmon (*Oncorhynchus nerka*) is perhaps a good choice for a model with synchronous oocyte development since a radioimmunoassay (RIA) is available for monitoring plasma gonadotrophin levels. Although a careful correlation with oocyte development has not been made, there is a trend towards increasing plasma gonadotrophin levels with ovarian recrudescence (see Peter and Hontela 1978). Because a RIA exists for assaying plasma gonadotrophin levels, and because ovarian development data have been published, the goldfish (*C. auratus*) could serve as the group-synchrony model. An increase of plasma gonadotrophin is not an essential aspect of ovarian recrudescence in goldfish (e.g. Peter and Hontela 1978). The magnitude and timing of the peak in the daily plasma gonadotrophin rhythm is significant with regards to ovarian development. Even though a gonadotrophin assay is not available presently, *F. heteroclitus* may make an ideal model for asynchrony because the dynamics of oocyte development are well documented. Furthermore, this species can be obtained in relatively large numbers, a considerable amount is known about its ecology and physiology, and it can be readily maintained in the laboratory.

Questions of possible phylogenetic, ecological, and geographic relationships of oocyte development patterns are seldom addressed. Among teleost fish, type of oocyte development cannot be obviously connected to any of these factors, nor are relationships between type of oocyte development and mating behaviours or parental care readily apparent.

Among Agnathans, lampreys show synchronous oocyte development and hagfish are characterised by serial development of follicles. Chondrostean and Holostean fish apparently have group-synchronous development. Synchronous oocyte development may be restricted to teleostean taxonomic groups which had an early evolutionary origin (e.g. Salmoniformes and Anguilliformes). In teleosts synchrony may be a highly specialised adaptation (see below) rather than an evolutionary remnant, since oocyte development among Chondrostean and Holostean fishes is group-synchronous. Synchronous oocyte development may be a highly specialised characteristic, possibly restricted to species which make long, energy-demanding migrations to spawning grounds. This type of oocyte development would not seem adaptive in highly variable environments or in habitats where food availability is high and conditions reproductively favourable for prolonged periods. Some salmon may not have recruitment stock remianing after a spawning, whereas other species have small oocytes or oogonia remaining after oviposition (Robertson and Wexler 1960); therefore, these fish basically may have group-synchronous ovaries with death after spawning one clutch.

Seasonally variable environments are prevalent on earth during the present geological period. Such environments require that preparations be made for the season which is predictably auspicious for reproduction. To be continuously prepared for reproduction is costly from an energetics standpoint, and probably not feasible in a variable environment due to changes in food availability. Developing a clutch of oocytes for this favourable season in advance (e.g. group synchrony) seems prudent. A prolonged period of conditions conducive for reproduction could be adapted to by the generation of a second or more clutches. The prevalence of group-synchrony may reflect the adaptability of this type of oocyte development. An intriguing, yet academic, concept is that group-synchrony provided the 'substrate' for the evolutionary emergence of other types of oocyte development. Although not an aspect of the definition of group-synchrony, a common characteristic of this type of oocyte development is no more than a single clutch simultaneously undergoing vitellogenesis. For one clutch of oocytes to be entering the phase of yolk accumulation while another clutch is completing the process is not uncommon. Reported cases of more than 1-2 distinct clutches in the vitellogenic phase bear careful documentation.

More thorough validation of asynchronous oocyte development is desirable, especially among fishes in tropical habitats. It is now clear that multiple spawns per season are not necessarily indicative of asynchrony nor can asynchrony be based on ovarian histological sections. Asynchronous oocyte development may be much less common than originally assumed. Although asynchrony has been established among cyprinodontids, claims for its existence outside this group are more tenuous. Among syngnathoids (e.g. seahorses), in which the ovary consists of a tubular sheet, oocytes of progressively greater size occur from one edge of the ovary to the other (e.g. Wallace and Selman 1981). Ovaries of these fishes have not been analysed carefully, yet the presumed asynchrony is rather divergent from that noted among cyprinodontids.

Asynchronous oocyte development apparently engenders the potential of opportunistic oviposition. Asynchrony predictably would be adaptive in environments where conditions are conducive for reproduction over a prolonged period, but also where short-term instabilities or fluctuations occur. This pattern of oocyte development is not generally manifested by species with short spawning seasons. Asynchrony lends itself to a small clutch size which can be oviposited opportunistically (i.e. this pattern may be an adaptation for releasing small numbers of eggs over a long spawning period). With asynchrony, spawning could be protracted in time and space to take advantage of favourable conditions. Additionally, there would be less pressure to mobilise substantial energy for a large reproductive effort, and a female could be less burdened by a large quantity of oocytes at one time. In an environment which has short-term fluctuations, one reproductive strategy is to be prepared for taking advantage of auspicious situations. Asynchrony can endow this preparedness, and loss of a few clutches would not be critical because of the relatively small investment per oviposition.

11 THE CONTROL OF TROUT REPRODUCTION: BASIC AND APPLIED RESEARCH ON HORMONES

A.P. Scott and J.P. Sumpter

This chapter covers three major topics of interest to endocrinologist and fish farmer alike: the hormonal control of the reproductive cycle, the timing of gamete production, and the control of sex. The second and third topics deal with practical matters which to a great extent depend upon a thorough understanding of the endocrinology of reproduction, which is the subject of the first topic. The rainbow trout (*Salmo gairdneri*) has been singled out for detailed consideration because of the wealth of knowledge now available for this species, and because of its importance to fish farming.

Hormonal Control of the Reproductive Cycle in Rainbow Trout

Females

The annual reproductive cycle of the female rainbow trout has been divided into four physiological stages (Bohemen and Lambert 1981), termed (1) previtellogenesis, (2) endogenous vitellogenesis, (3) exogenous vitellogenesis, and (4) ovulation and spawning. Much effort has been directed within the last four years to describing changes occurring during these stages: changes in ovarian histology (van den Hurk and Peute 1979), pituitary histology (Olivereau 1978; Peute, Goos, Bruyn and van Oordt 1978; van Putten, Peute, van Oordt, Goos and Breton 1981), steroid biosynthesis (Lambert and van Bohemen 1979; Sire and Dépêche 1981; van Bohemen and Lambert 1981), plasma sex steroid levels (Billard, Breton, Fostier, Jalabert and Weil 1978; Lambert, Bosman, Hurk and van Oordt 1978; Whitehead, Bromage and Forster 1978; Scott, Bye and Baynes 1980; Scott, Sheldrick and Flint 1982; Scott, Sumpter and Hardiman 1983; Campbell, Fostier, Jalabert and Truscott 1980; van Bohemen and Lambert 1981), plasma vitellogenin levels (Whitehead *et al.* 1978; Campbell and Idler 1980; van Bohemen, Lambert and Peute

1981; Sumpter 1983) and plasma and pituitary gonadotrophin levels (Billard *et al.* 1978; Fostier, Weil, Terqui, Breton and Jalabert 1978; Jalabert and Breton 1980; van Putten *et al.* 1981; Bromage, Whitehead and Breton 1982). The levels of hormones and vitellogenin (see Chapter 10) in the blood show up especially clearly the various stages of the reproductive cycle. Figure 11.1, for example, shows the plasma levels of testosterone, oestradiol-17β, 17α-hydroxy-20β-dihydroprogesterone, vitellogenin and gonadotrophin in females of a January-spawning strain of rainbow trout, and it can be seen that they undergo marked changes throughout the year.

During the first physiological stage of the cycle (previtellogenesis) the reproductive system appears to be quiescent, as indicated by the very low levels of sex steroids in the plasma. However, before it is concluded that this period is one of endocrine inactivity, more thorough studies must be conducted. For instance, the selection of oocytes that will develop into eggs during the next reproductive season may occur during this stage.

During the next stage, that of endogenous vitellogenesis, the oocytes themselves actively synthesize a proteinaceous material which they store within yolk vesicles (summarized by Wallace and Selman 1981). The nature and significance of this material within the yolk vesicles remains unclear; presently there is no evidence that it bears any relationship to the exogenously produced vitellogenin. These yolk vesicles fill the oocyte during the early stages of its development, but are pushed to the periphery during the later stages of vitellogenesis, when vitellogenin is sequestered from the blood. It is also not known whether exogenous vitellogenesis necessarily succeeds endogenous vitellogenesis, or whether the two processes might not overlap. Sensitive radioimmunoassays reveal that the level of vitellogenin in plasma is steadily increasing during the period of endogenous vitellogenesis (Figure 11.2). Since oestradiol -17β levels, which regulate exogenous vitellogenin production by the liver, are also significantly raised at this time, it appears that exogenous vitellogenin production begins much earlier than had previously been realised.

The control of endogenous vitellogenesis is also not understood. It may be under the control of gonadotrophin, as some studies have detected elevated levels of gonadotrophin in plasma of female brown trout (Billard *et al.* 1978) and rainbow trout (Figure 11.3) during this period. We, however, have failed to record elevated gonadotrophin levels at the same stage (Figure 11.1). There are a number of possible

Figure 11.1: Annual changes in plasma levels of sex steroids, vitellogenin and gonadotrophin in females of a winter-spawning strain of rainbow trout. All levels were measured by radioimmunoassay. Bars (as shown in Figures 11.1-11.6) represent standard error of mean (n = 7—12). 17α, 20βP = 17α-hydroxy-20β-dihydroprogesterone.

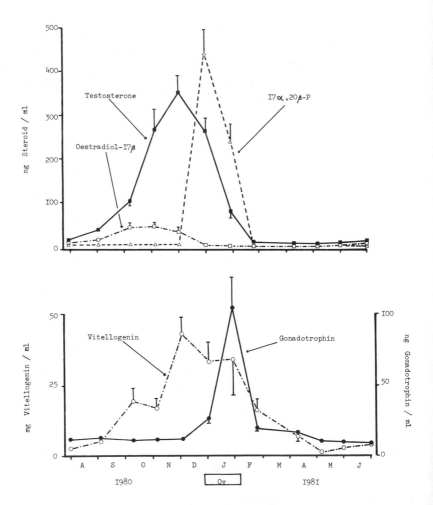

Figure 11.2: Early changes in levels of sex steroids and vitellogenin in 3-year-old females of an autumn-spawning and winter-spawning strain of rainbow trout.

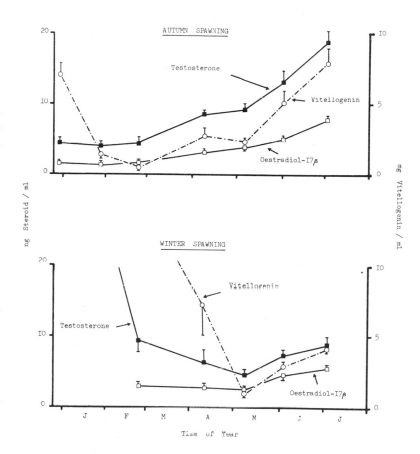

reasons for this difference. First, the number, identity and chemical structure of the gonadotrophin(s) in salmonid pituitaries are still uncertain. Idler and his colleagues (1975) have isolated distinct vitellogenic and ovulatory gonadotrophins from chum salmon pituitaries, though others have reported the existence of only one gonadotrophin (this aspect is fully discussed in Dodd and Sumpter 1983). This uncertainty makes it impossible to know exactly what the existing salmonid gonadotrophin radioimmunoassays measure.

Figure 11.3: Changes in serum immunoreactive gonadotrophin levels in female rainbow trout maintained under each of the following light regimes:
●—● Control, 12 month seasonal cycle; ○—○ Seasonal cycle compressed into 9 months; ■—■ Seasonal cycle compressed into 6 months. The water temperature was a constant 9°C throughout. Spawning occurred in mid-January under the 12 month cycle, early December under the 9 month cycle and mid-October with the 6 month cycle.

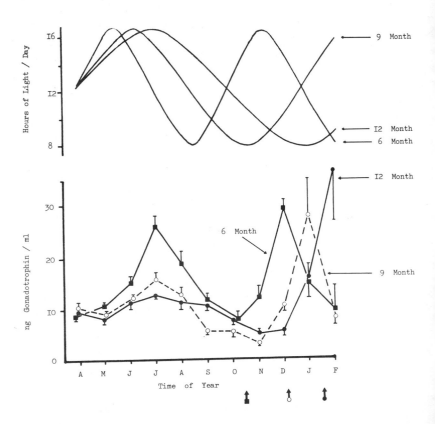

Source: Whitehead, Bromage and Breton, unpubl. obs.

Secondly, it is possible that the existing gonadotrophin radio-immunoassays may also be detecting circulating thyroid-stimulating hormone, which has a similar structure to gonadotrophin. Thirdly, release of gonadotrophin during the early part of the reproductive cycle may be episodic (Zohar 1980), and results may thus be affected by the timing and frequency of blood sampling. At the moment, blood gonadotrophin measurements (other than those accompanying ovulation; see later) should be interpreted with care.

The lack of agreement between the different studies highlights the fact that the present state of our knowledge concerning gonadotrophin(s) in salmonid pituitaries is still extremely unsatisfactory. This is, we feel, an area of high priority for research.

The third stage, of exogenous vitellogenesis, is characterised by high levels of oestradiol-17β (40-50 ng ml^{-1}), oestrone (van Bohemen and Lambert 1981), testosterone and vitellogenin. Several earlier studies have reported considerably lower levels of oestradiol-17β in female rainbow trout at this stage (4-5 ng ml^{-1}) (Schreck, Lackey and Hopwood 1973; Whitehead, Bromage and Forster 1978; Whitehead *et al.* 1978). In our experience, when using volumes of fish plasma greater than 50 μl in assays, there is considerable interference (even after extraction of steroids with diethyl ether), such that oestradiol-17β levels appear to be lower than they should be. This may be the reason for the large disparity in reported levels. The results shown in Figure 11.1 have been confirmed by a more specific assay (Scott, Bye and Baynes 1980), and are similar to the levels reported by van Bohemen and Lambert (1981). A primary function of oestradiol-17β, and possibly oestrone, is the stimulation of vitellogenin production by the liver (reviewed in Dodd and Sumpter 1983). The function of testosterone in females is less clear. There are several possibilities. Firstly, as the probable precursor of oestradiol-17β, it might be released incidentally into the circulation and have no primary function. Secondly, since it has been shown that injections of both testosterone and oestradiol-17β will stimulate gonadotrophin synthesis by the pituitary gland of immature rainbow trout (Crim *et al.* 1981), and also that some peripheral tissues such as the brain (but not the liver) readily convert testosterone to oestradiol-17β (Lambert and van Bohemen 1980) it may be considered to be a target-organ-specific oestrogen. Thirdly, it is possible that the plasma testosterone levels are high because the steroidogenic cells of the ovary are required to be superactive at the time of oocyte maturation. Plasma levels of the 17α-hydroxylated progestagens rise extremely rapidly before ovulation (Figure 11.4), and we suggest (see below) that this occurs as a result of the blockade of ovarian androgen synthesis (indicated by the fall in testosterone levels (Figure 11.4), rather than by *de novo* synthesis.

During the period of exogenous vitellogenesis the vitellogenin secreted by the liver under oestrogen stimulation is accumulated in the developing oocytes, which store it as yolk. Over a few months the ovary grows until, when fully developed, it weighs up to one quarter of the weight of the whole fish. This quite phenomenal growth requires the incorporation of very large quantities of yolk into the oocytes, which is made possible by the secretion of vitellogenin from the liver at such a rate that blood levels reach tens of milligrams per millilitre.

The final stage of the reproductive cycle, comprising oocyte maturation, ovulation and the immediate post-ovulatory period, has been well described in salmonids. In rainbow trout the period from the beginning of oocyte maturation (when meiosis is resumed) to ovulation (when the oocytes are released into the body cavity) takes about 50 degree (°C) days (that is, at 10°C this would take 5 days) (Bry 1981). There is general agreement that secretion of a gonadotrophin by the pituitary is a prerequisite for oocyte maturation and ovulation and that the action of this GTH on oocyte maturation is probably mediated through the steroids 17α-hydroxy-20β-dihydroprogesterone and 17α-hydroxyprogesterone (Scott, Sumpter and Hardiman 1983). We have recently measured plasma levels of these hormones (and also testosterone and oestradiol-17β) during the complete sequence of events from oocyte maturation to the stage of ovarian quiescence in female rainbow trout (Figure 11.4; Scott, Sumpter and Hardiman 1983). One interpretation of these results is that there is feedback inhibition between the sex steroids and GTH and that falling plasma oestradiol-17β levels allow an initial rise of GTH which stimulates the rapid rise in progestagen formation with a concomitant fall in androgen levels. The large post-ovulatory peak in plasma gonadotrophin levels, also recorded by Jalabert and Breton (1980), may be the result of a reduction in feedback inhibition of the sex steroids on the pituitary.

Males

There are several published studies on plasma androgen levels (Schreck 1972; Sanchez-Rodriguez, Escaffre, Marlot and Reinaud 1978; Whitehead, Bromage, Breton and Matty 1979; Scott, Bye,

Figure 11.4: Hormone changes during ovulation in female rainbow trout. Ovulation (day 0) was determined as the time that eggs could be freely expressed from the fish (n = 14). (From Scott, Sumpter and Hardiman 1983) (17α-20βP = 17α-hydroxy-20β-dihydroprogesterone; 17α-P = 17α-hydroxyprogesterone).

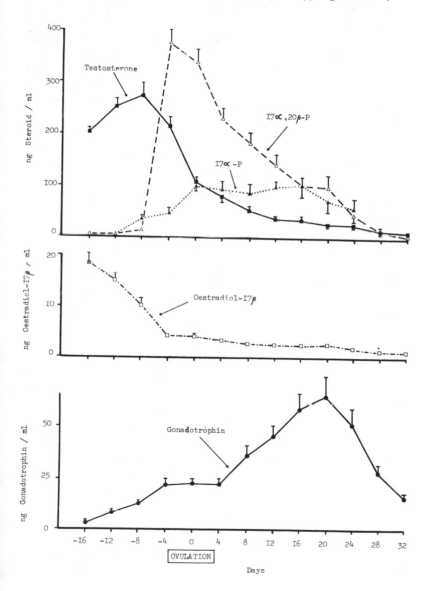

Baynes and Springate 1980), testicular histology (Billard *et al.* 1978) and sperm yield (Billard, Breton and Jalabert 1971; Sanchez-Rodriguez *et al.* 1978) of male rainbow trout throughout the year. It is apparent that there are well-defined periods of quiescence, spermatogenesis and spermiation (Figure 11.5).

Major androgens that have been identified in trout plasma are testosterone, 11-ketotestosterone (11-KT) and testosterone glucuronide (Kime 1980). The peak levels of 11-ketotestosterone were shown to occur later in the season than the peak levels of testosterone by Scott *et al.* (1980), and it was suggested that 11-ketotestosterone may be involved with spermiation. Further studies by Baynes and Scott (unpubl. obs.), however, indicate that in some strains of trout the 11-ketotestosterone peak coincides with that of testosterone, and is unrelated to peak sperm production which occurs well after the peak levels of both androgens (Figure 11.5). Presumably both androgens control other aspects of testicular growth, but whether they are necessary for all of spermatogenesis or to regulate specific stages only, is unclear. The development of secondary sexual characteristics in salmonids also appears to be attributable to 11-ketotestosterone, e.g. skin coloration and kype growth (Idler, Bitners and Schmidt 1961). The function of 17α-hydroxy-20β-dihydroprogesterone which was first identified in plasma of male Pacific salmon in 1962 (Schmidt and Idler 1962) and which appears to be present in plasma of spermiating fish only (Figure 11.5), is unclear.

In the male trout, as in the female, the secretion of the gonadal hormones is thought to be under the control of gonadotrophin(s) (Billard 1978), though the practical difficulties associated with the measurement of blood gonadotrophin levels in salmonids (see above) have precluded an accurate assessment of their involvement in testicular development.

The Timing of Gamete Production

The timing of gamete production is a subject of considerable importance to fish farming. Because of the seasonal nature of egg production, hatcheries are under utilised for most of the year and methods for altering the timing of gamete production enable their year-round use, and also provide consistent supplies of eggs and fingerlings for the fish farmer.

Figure 11.5: Changes in sex steroids and sperm yield of male rainbow trout over a complete annual cycle (n = 10—15) (from data by S.M. Baynes, Lowestoft, UK, in prep.) (17α-20β-P = 17α-hydroxy-20β-dihydroprogesterone).

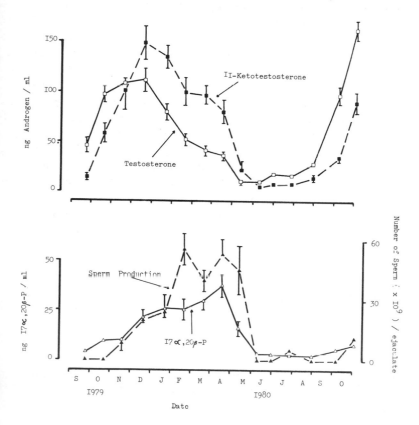

There are several ways of achieving this aim: photoperiod manipulation, hormone injections, genetic selection for spawning time, cryopreservation of gametes and controlling the rate of embryonic development.

Environmental Control of the Reproductive Cycle

Like the vast majority of plants and animals living at mid- and high latitudes, salmonids do not breed continuously, but confine their breeding to that time of the year when the offspring have the greatest chance of survival. The normal time at which particular species spawn varies quite considerably in salmonids. Pacific salmon, brook and brown trout normally spawn in the autumn (September—November), and the original 'Shasta' strain of rainbow trout (from which farm stocks were derived) in early spring (February—March) (Frost 1974). Cultivation of the rainbow trout has resulted in numerous strains with a wide range of spawning times (Busack and Gall 1980). In order to be able to spawn at the correct time, the trout needs to be able to identify what time of year it is. There are two ways of achieving this; one is to have an endogenous clock with a period of close to one year (a so-called circannual rhythm), the other is to evolve the ability to monitor some aspect of the environment that undergoes yearly cycles. Obviously, there will be intense selective pressure to utilise a factor that is constant from year to year, rather than one that is less predictable. To this end the vast majority of plants and animals, including rainbow trout, have chosen daylength (that is, the ratio of light to dark per day) as a major factor controlling the time at which they breed. The mechanics of this photoperiodism have been of interest to both scientists and fish farmers since the pioneering studies of Hoover and Hubbard (1937) on the brook trout, *Salvelinus fontinalis*. Photoperiod control of spawning in rainbow trout was first investigated by Japanese workers in the early sixties (Shiraishi and Fukuda 1966; Nomura 1962) and after a considerable interval by French (Breton and Billard 1977) and English groups (Whitehead *et al.* 1978).

Following Hoover and Hubbard's (1937) demonstration that the spawning time of brook trout could be advanced 3 months by compressing the yearly cycle of daylength, a similar procedure was shown to advance the spawning time of rainbow trout (Nomura 1962; Kunesh, Freshman, Hoehm and Nordin 1974; Whitehead *et al.* 1978; Figure 11.3). The procedure has been applied on commercial farms to obtain out-of-season spawning of brook trout (Corson 1955) and rainbow trout (Buss 1980), but it has proved to be too complicated in practice. Efforts have thus been directed in recent years to unravelling the mechanism of action of photoperiodism in salmonids, with the aim of producing a simpler procedure for the fish farmer.

It has long been assumed that gonad growth in salmonids is triggered by the decreasing day lengths of the late summer and autumn (Whitehead *et al.* 1978; Billard, Bry and Gillet 1981). The main evidence in support of this contention is that the marked increases which occur in the size of the gonad and also the elevations of plasma sex steroid levels are coincident with the late summer and autumn months in all salmonids (even gonad growth in fish subjected to compressed light cycles tends to coincide with the decreasing light phase). Abrupt changes of daylength from constant long (e.g. eighteen hours of light followed by six hours of darkness each day; 18L:6D) to constant short days (e.g. 6L:18D) were also shown to be effective in accelerating gonadal development (Whitehead and Bromage 1980). Additionally, many of the early experiments showed that withholding the period of shortening day lengths (by giving the fish additional light after the summer solstice) usually delayed the spawning season by 3-6 weeks (Allison 1951; Hazard and Eddy 1951; Eriksson and Lundquist 1980).

The results of more recent research, however, do not support this assumption, but point to the contrary fact that it is the long day lengths encountered by fish in the spring and early summer which initiate gonad recrudescence. In brown trout and rainbow trout it is clear, for example, when sensitive radioimmunoassays are used to monitor indices of gonad development, that under natural photo-period conditions, plasma levels of gonadotrophin (Billard *et al.* 1978; Bromage, Whitehead and Breton 1982), oestradiol-17β (Billard *et al.* 1978; Figure 11.2) testosterone and vitellogenin (Figure 11.2) start to rise long before the summer solstice. Indeed, in autumn-spawning rainbow trout, sex steroid and vitellogenin synthesis appear to begin afresh as early as March. Further evidence is provided by the fact that: (1) ovarian recrudescence in brook trout and rainbow trout is stimulated by a long constant photoperiod regime (18L:6D) if it is applied early in the year (Henderson 1963; Whitehead and Bromage 1980); (2) delaying the onset of long photoperiods (e.g. by holding fish on a 6L:18D photoperiod regime from March to June) will appreciably delay the spawning time of female rainbow trout (Shiraishi and Fukuda 1966); (3) male rainbow trout held under constant 18L:6D from the fingerling stage mature prematurely at 1 year old (Skarphedinnson, pers. comm.). Collec-tively, these results suggest that it is the lengthening days of spring which may initiate gonad development in the salmonids.

Presently, the best interpretation of all the experimental data on photoperiodic control of reproduction in salmonids is that the lengthening days of late winter and spring probably initiate sexual development, while the shortening days of later summer and autumn accelerate gonad growth. At the moment it is difficult to give a more adequate explanation of photoperiod control in salmonids and there are still many results that cannot be fitted into a single theory. Pyle (1969), Potson and Livingstone (1971) and Henderson (1963), for example, have all observed that male fish often mature earlier in the year than females in response to experimental photoperiods. (This is a separate phenomenon from that of premature maturation, where male rainbow trout are physiologically able to mature, and often do mature, at 1 year old, while females cannot mature until they are at least 2 years old (Purdom 1979).)

Another problem arises with strain variation. For example, we found (unpubl. obs.) that females of two strains of rainbow trout with different natural spawning times (Figure 11.2) had a completely different response to a continuous long photoperiod (18L:6D) applied early in the year (March) — the spawning time of October/November spawners was delayed and the spawning time of January/February spawners was advanced. This evidence strongly suggests that in the two strains of rainbow trout different daylengths are required to initiate maturation. It also suggests that from the practical point of view, photoperiod regimes may have to be tailored to a farmer's particular broodstock. Another intriguing observation was that 2-year-old female rainbow trout held under a constant 18L:6D for 18 months exhibited a reproductive cycle of roughly 6 months (Figure 11.6). This phenomenon has also been observed in the brook trout, *Salvelinus fontinalis* (Henderson 1963), stickleback, *Gasterosteus aculeatus* (Baggerman 1980), and dab, *Limanda limanda* (Htun-Han 1975). While its physiological significance is obscure, it may be open to practical exploitation.

The age of the fish may also influence their response to a particular photoperiod. Many experimenters have used 2-year-old immature fish (e.g. Henderson, 1963; Whitehead *et al.* 1978). While the use of 2-year-old females may simplify problems for the scientist, who can be certain that none of the fish will have spawned the previous year, their use on a commercial fish farm is unlikely to be welcome due to the small eggs that are produced by first-time spawners (Kato 1975). Small eggs generally have a lower hatching rate and produce less viable and slower-growing larvae than large eggs (Gall 1974). It has also been reported that fish induced to mature early by photoperiod manipulation yield smaller eggs than normal fish from the same stock (Nomura 1962; Buss 1980). Since it is necessary that out-of-season eggs should be of an acceptable size to fish farmers, it would seem logical to commence a photoperiod regime with mature 3-year-old females (the age of the males is probably of no importance). It is clear,

Figure 11.6: Plasma levels of testosterone in female trout (n = 5 up to month 13, and then n = 2) held under a constant 18L:6D photoperiod. The regime was started in March using 2-year-old females of a January-spawning strain. Temperatures varied from 8—14°C. The timing of the first ovulation in individual fish was not synchronised (3 spawned in July, 1 in October and 1 in December), but the data have been plotted to show that the second ovulation in all fish (and third in two fish) occurred after only a six-month interval.

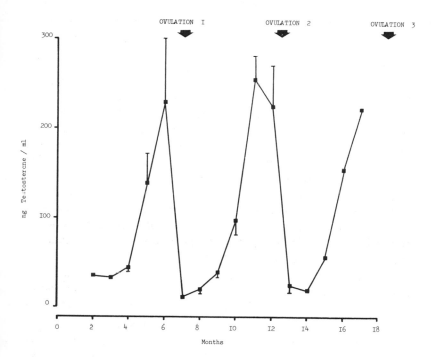

however, that the reproductive history of 3-year-old females (i.e. whether the fish spawned at 2 years old or not) can affect their response to photoperiod (Henderson 1963). Dodd and Sumpter (1982) stress that this factor is of major importance in interpreting photoperiod experiments carried out on vertebrates.

The influence of temperature on photoperiodic control of reproduction in salmonids has received little attention. Most experimental work has been carried out using water of a constant temperature.

However, many trout farms are supplied by river water of a fluctuating temperature which, in the UK for example, can reach 19°C in late summer and 1-2°C in winter. It is necessary, therefore, to know how temperature fluctuations may affect the response to photoperiod. There is limited evidence that temperature can have a direct effect. Henderson (1963) found that female brook trout at 16°C responded positively to a long photoperiod while fish at 8°C did not. There is no evidence, however, that temperature changes act as triggers of gonad maturation in salmonids as they do in cyprinids (Billard *et al.* 1978). Photoperiod acceleration of spawning has been obtained with unsynchronised water temperature cycles in *Oncorhynchus* spp. (MacQuarrie, Markert and Vanstone 1978). There is, however, a major problem with inducing spawning in conditions where the water temperatures exceed 14-15°C. Ovulated eggs will deteriorate rapidly in females at these temperatures (Escaffre, Petit and Billard 1977), and spermiation may be reduced or even inhibited in males (Billard and Breton 1977).

Although the photoperiodic cycle appears to be the primary proximate cue timing the breeding cycle of trout, fine adjustments might be made using other factors, e.g. nutrition, water quality, the presence of individuals of the opposite sex, etc. Very little experimental information is available on these ultimate cues. Circumstantial evidence for their existence is provided, however, by observations such as the inability of most wild rainbow trout to spawn naturally in the rivers and lakes in Europe, even though the fish undergo complete gonadal development (Frost 1974). Presumably some ultimate cue, possibly the appropriate substrate for a redd, is lacking, and this prevents oviposition.

The role of endogenous circannual rhythms in regulating the timing of fish reproduction is unclear, though they may be more important than has previously been realized. Whitehead *et al.* (1978) maintained female rainbow trout on a constant photoperiod (12L:12D), with the water temperature at 9°C throughout, for a year, and showed that these fish completed a reproductive cycle, and spawned at the same time, as the control fish maintained under natural environmental conditions. Though there are a number of difficulties in interpreting this type of experiment, which are fully discussed in Dodd and Sumpter (1983), one interpretation is that the fish were able to breed at the correct time of year without needing to perceive any changing environmental information.

It seems clear that, for the moment, the development of practical systems for photoperiod control of reproduction will have to be carried out empirically, as our understanding of the basic control mechanisms of photoperiod in salmonids is inadequate. When, in addition to the uncertainty of our knowledge, the variability of conditions, broodstocks, general management and objectives on commercial farms is taken into account, the prospect of setting up a successful photoperiod system can be quite daunting.

Hormonal Manipulation of the Reproductive Cycle

Regulation of Early Phases. The enhancement or acceleration of vitellogenesis and gonad recrudescence may also be achieved by hormone injection (Lam 1982). Donaldson and his co-workers (reviewed by Lam 1982) induced precocious sexual maturation in male and female pink salmon (*Oncorhynchus gorbuscha*) by regular injections of pituitary extracts. The frequent manipulations entailed by such a method, however, would make it of limited value for a commercial fish farm. Furthermore, much of the work concerned with hormonal manipulation of the early phases of the reproductive cycle (Crim, Peter and Billard 1981) suggests that there is a sex difference in response, with males being much more likely to respond than females. Overall, until our understanding of the endocrine control of the reproductive cycle improves considerably, there seems little prospect of hormonal manipulation of its early phases becoming a useful technique on the fish farm.

Induced Ovulation in Female Rainbow Trout. Fortunately, in farmed rainbow trout broodstocks and most other salmonids, the processes of oocyte maturation and ovulation occur spontaneously. The development of a procedure to induce ovulation artificially would therefore appear to be unnecessary. However, Jalabert, Breton and Bry (1980) have suggested that artificial induction could be used by fish farmers to synchronise the ovulation time of rainbow trout. Normally, less than 20 per cent of females in a single broodstock will be mature at any time over a normal spawning period of 6-8 weeks (Kato 1975). This usually means that all females in a broodstock must be handled regularly (usually once a week) until they are found to be ready for egg stripping.

Donaldson and his co-workers (see review by Lam 1982) have investigated a variety of techniques for inducing ovulation in salmonids. Injections of hormones acting at the hypothalamic-pituitary level (e.g. LHRH) and at the ovarian level (e.g. gonadotrophin and 17α-hydroxy-20β-dihydroprogesterone) and also 'endocrine active' drugs (e.g. the anti-oestrogen Tamoxifen) have all been shown to accelerate oocyte maturation and ovulation in salmonids. All of these substances work best when a small priming dose of gonadotrophin or salmon pituitary extract is given to the fish 2-3 days prior to their injection (Donaldson, Hunter and Dye 1981). Perhaps the simplest procedure, and one which has been well validated (Hunter, Donaldson and Dye 1981; Scott, Sheldrick and Flint 1982) is to inject extracts of acetone-dried chum salmon pituitary powder at two dose levels. Whether this procedure would be taken up on commercial fish farms is doubtful. There is firstly the problem of the parenteral administration of substances to fish which must be carried out by a qualified person (e.g. veterinary surgeon or animal scientist). There is also some doubt about the quality of the eggs produced by artificial induction (Bry 1981). Finally, there is considerable doubt as to whether most farmers would consider it an advantage to be able either to synchronise the ovulation times of their broodstock or obtain eggs only 3-4 weeks earlier than normal.

Artificial Stimulation of Spermiation.   While the females of a particular strain of rainbow trout will spawn over a period of only 6-8 weeks, males can spermiate for as long as 6 months (Figure 11.5), and since very little milt is needed to fertilise eggs (see review by Scott and Baynes 1980) there is little incentive to stimulate spermiation artificially. It can probably be achieved, however, by injections of pituitary hormones or progestagens. These are effective in other species (de Montalambert, Bry and Billard 1978).

Genetic Manipulation of Spawning Time

The time of spawning in rainbow trout has a strong genetic component (Purdom 1979). The Americans, for example, have achieved a remarkable spread in the timing of egg production of rainbow trout by selection for early and late spawning characteristics (Busack and Gall 1980). Such work can only be carried out in the larger hatcheries, as it involves careful management and separate accommodation for several broodstocks. Nevertheless, in the long run, it is probably going to be the most satisfactory method of extending the egg season.

Within the UK, the availability of strains having different spawning times is limited (Scott 1981). Whether this can be rectified by importing certain strains from the USA, or by a new programme of selection, remains to be determined. From 1976 to 1980, partly due to the difficulty in obtaining eggs at certain times of the year, approximately 20 per cent of the eggs required for table trout production in England and Wales were imported from the USA and Denmark (Scott 1981). Analysis of the import figures revealed that the American eyed eggs arrived mainly in November and Danish eyed eggs in March and May (Scott 1981). The evidence indicates that, at the moment, this is still the most popular way of extending hatchery operations in the UK. Prior to 1975, a limited number of eggs were imported from the southern hemisphere (i.e. eggs that were regularly available in the middle of summer).

Cryopreservation of Gametes

Gamete preservation, besides being another way of controlling the timing of hatchery operations, has the added advantages of reducing the number of males kept as broodstock, enabling gametes to be transported from one hatchery to another, and allowing the banking of unusual and useful genetic traits. Although the cryopreservation of eggs is not yet possible (Erdahl and Graham 1980), the cryopreservation of sperm is relatively easy (Scott and Baynes 1980). However, although many workers have now achieved high fertilisation rates with specific hatches of cryopreserved rainbow trout sperm, replication of results is still a problem. Until this is solved, cryopreservation techniques cannot be applied to large-scale hatchery operations. They are, however, quite adequate for gene banking.

Prolonging Egg Incubation

The direct effect of temperature on the rate of embryonic development can be used to extend the egg season by 2-3 months. This is because the time taken from fertilisation to hatching in rainbow trout is related exponentially to water temperature over the range 3.1-16.9°C. Kawajiri (1927) derived the equation $\log T = 2.27 - (0.068) \times \Theta$, where T = time in days and Θ = water temperature in °C. From this equation it can be seen that eggs would take 160 days to hatch at 1°C, 100 days at 4°C and only 39 days at 10°C.

Erdahl and Graham (1980) have explored the possibility of cooling eggs to sub-zero temperatures in the presence of a cryoprotectant: which effectively halts development, but does not freeze the egg contents. Their results were extremely promising, and suggest yet another useful way in which hatchery operations can be controlled.

The Control of Sex

The third major topic in which endocrinology has played a direct role is that of sex control.

For the fish farmer, sexual development in his stocks can create several problems (Bye and Jones 1981; Bye and Lincoln 1981), the major one being that a significant proportion of males will generally mature in their first year, before the stock can be marketed; the secondary sexual characteristics associated with maturity in males reduce disease resistance and cause a deterioration in appearance and flesh quality which can decrease their sale value considerably. Two approaches have been taken to resolve this problem: (1) to produce females only, and (2) to produce sterile fish.

Producing All-female Trout

It has been established that only male trout have the potential to produce male offspring (i.e. male trout are heterogametic (possessing a Y chromosome) (Bye and Jones 1981). Although trout sex is normally determined by the presence or absence of the Y chromosome, endocrinologists have shown that the differentiation of the gonads can be influenced by feeding fry with diets containing low levels of sex steroids: oestradiol-17β will promote the formation of an ovary, and methyltestosterone the formation of a testis (see review by Donaldson and Hunter 1982). While direct feminization with oestradiol-17β would seem the easiest way to obtain all-female stocks, it has proved unreliable and troublesome (Bye and Lincoln 1981). An indirect technique, initially involving the masculinization of females, has, on the other hand, proved easy to perform and control, and has been taken up enthusiastically by the trout farming industry in the UK (Bye and Lincoln 1981). The procedure used in the UK to convert genetically female fish into functional males is to administer 3 mg kg^{-1} 17α-methyltestosterone in the diet of first-feeding fry for about 75 days at about 10°C (Bye and Lincoln 1981). The milt produced by adult masculinised females (which can be easily distinguished from the ordinary males by the absence of a Wolffian duct (see Scott *et al.* 1980) is used to fertilise normal eggs, and since no Y chromosomes are present in any of the sperm, female-only fish will be produced.

There are other less reliable ways of producing female-only fish. Since female trout are homogametic it is possible to use diploid gynogenesis, which involves the fertilisation of eggs with UV or γirradiated sperm, and a subsequent temperature shock to double-up the chromosome number (Chourrot 1980; Donaldson and Hunter 1982). This method, however, results in an increased degree of inbreeding, and is often accompanied by heavy egg mortalities (Chourrot 1980). There is one other possibility — to sort the sexes manually at an early stage. However, this is virtually impossible in rainbow trout since, until they are nearly mature, there is no easy way of distinguishing sex. Several biochemical methods can be applied when the gonads are maturing (i.e. up to 6 months before the spawning season). Measurements can be made, for example of vitellogenin (Le Bail and Breton 1981), 11-ketotestosterone (Sangalang, Freeman and Fleming 1978) or oestradiol-17β, all of which are likely to be much higher in one sex than the other.

Producing Sterile Trout

Sterile fish may be more useful than all-female fish in certain situations.

The most promising method of producing sterility at the moment is by induction of triploidy — preventing the dissociation of the second polar body during the first hour of egg development. This has been achieved in a number of species by subjecting the newly-fertilised eggs to a temperature shock (Purdom 1972). Until very recently attempts to induce triploidy in salmonids, mainly by 'cold' shocks, were unsuccessful (e.g. Lincoln, Aulstad and Grammeltveldt 1974), but Chourrot (1980) showed that high levels of triploidy could be achieved in rainbow trout using mild 'heat' shocks soon after fertilisation.

Although triploidy is responsible for impaired gonad development in fish it has recently been shown that the degree of impairment differs between the sexes. In triploid male plaice (*Pleuronectes platessa*) and in the hybrid with the flounder (*Platichthys flesus*) (Lincoln 1981a), for example, there is little or no reduction in testis growth compared with diploid controls. Mature triploid females, on the other hand, had ovaries which were less than 5 per cent of the weight of those in control diploid fish (Lincoln 1981b). Thorgaard and Gall (1979) found that in naturally-occurring rainbow trout triploids, the females also showed virtually no gonad development, while the males were morphologically indistinguishable from diploid males.

On the basis that there is likely to be little commercial advantage in producing male triploid rainbow trout due to this lack of suppression of gonad growth, Lincoln and Scott (1983) have exploited the two techniques — one for producing all-female progeny (using sperm from masculinised females), and the other for producing triploids (using heat shocks) — to produce batches of sterile all-female triploids for commercial trials.

Laird and her co-workers (Laird, Wilson and Holliday 1981) have attempted to produce sterile fish by injection of minced gonads (to induce autoimmune rejection) but have had little success. Brown and Richards (1979) have described a technique by which gonads can be surgically removed from immature fish. Problems with this technique, however, are that the testicular tissue may regenerate, and only small numbers of fish can be treated.

Certain generic hybrids within the salmonids are sterile (Chevassus 1979). These however are unlikely to find a market in the UK, where tastes in fish are very conservative (Lewis 1980).

A purely endocrinological method for inducing sterility is to feed young fish, which have relatively undifferentiated gonads, a diet containing high concentrations of sex steroids (e.g. 25 mg kg^{-1} 17α-methyltestosterone, cf. 3mg kg^{-1} for masculinisation of trout); (Bye and Jones 1981; Donaldson and Hunter 1982).

Summary

We have attempted to describe three main areas of research, involving the endocrine basis of trout reproduction, the timing of gamete production, and control of sex, which we consider will lead to practical advances in trout farming. Some of the techniques described in the latter two areas are already operating on commercial farms in the UK, and the basic research being carried out on the endocrine control of gonad development will undoubtedly lead to practical advances in the future. Without the knowledge gained from this latter research we cannot hope to answer some of the as yet unanswered, but important, questions such as why sex steroids influence gonad differentiation, why male fish mature earlier than females and what factors determine fecundity, egg size and quality.

12 CONTROL OF REPRODUCTION IN ELASMOBRANCH FISHES

J.M. Dodd, M.H.I. Dodd and R.T. Duggan

Cartilaginous fishes have probably been in existence for well over 400 million years (Schaeffer and Williams 1977) and extant species comprise an important element in the contemporary fish fauna. The group may be subdivided into two Classes (or subClasses) of very unequal size, Elasmobranchii and Holocephali (sometimes the fossil Placodermi are included as a third subClass). Little is known about the latter (see Wourms 1977; Dodd and Sumpter 1983) and they will not be considered here. Living elasmobranchs are classified into 144-56 genera with some 739-803 species (Compagno 1977). These fall clearly into two distinct orders: Squaliformes (sharks and dogfishes) and Raiiformes (skates and rays). In view of their long evolutionary history, itself a testimony to the continuing success of their reproductive strategies, and their contemporary importance, it is unfortunate that they have received so little attention in the context of reproductive endocrinology. They have a full complement of the endocrine organs and tissues associated with reproductive regulation throughout the vertebrates, including a pituitary gland notable for its subdivision into regions, one of which has a special relationship with reproduction. They possess a differentiated hypothalamus and median eminence; a thyroid gland that is discrete and can be surgically removed, unlike the situation in most teleosts; and a pineal organ. Their reproductive endocrinology appears to be basically similar to that of other vertebrates, but there are interesting differences. Most elasmobranchs are viviparous, though in most cases the dependence of the developing young on the mother is slight. However, in some a placenta develops early in gestation and it would be particularly interesting to know whether the pituitary and ovary are involved in the maintenance of pregnancy in such cases. Most of what is known is restricted to descriptions of general reproductive biology, life cycles and general behaviour and to morphological, histological and histochemical aspects of reproductive structures. Physiological studies are virtually restricted to two squaliform species, the lesser spotted dogfish, *Scyliorhinus canicula* and the

221

spiny dogfish, *Squalus acanthias.* In spite of this, and to avoid tedious repetition of phrases like 'in the few species that have been investigated' we have made several generalisations which will have to suffice until many more species have been studied.

The Gonads and Sex-determination

The elasmobranch gonad, like that of other vertebrates, has a dual origin, its somatic tissues developing from the mesodermal genital ridges, whereas the germinal components, primordial germ cells or gonia, arise in the endoderm of the gut. The genital ridges in elasmobranchs, as in amphibians and amniotes, but not in teleosts, are differentiated into two regions, cortex and medulla. The gonia migrate from the gut to the genital ridges at a very early stage of development and whether they settle in the cortex or the medulla depends on their genotype: if the former, an ovary develops and if the latter, a testis. Sex is therefore genetically determined.

Experimental treatment of early embryos at the time of sex differentiation including immersion in oestrogens or androgens, or injection of these steroids into the yolk sac and transplantation of developing gonads into embryos of the opposite sex, have shown that the sex-determining mechanism is more stable than in many other vertebrates (Dodd 1960, 1983; Chieffi 1967).

In most vertebrates, oogenesis is cyclical and continuous, a stock of oogonia persisting throughout adult life, but in elasmobranchs, oogonial mitosis is completed before the onset of sexual maturity and the finite crop of oogonia that results enters the prophase of the first meiotic division precociously, the mature ovary containing only oocytes (Franchi, Mandl and Zuckerman 1962).

Modes and Patterns of Reproduction

A recent analysis by Wourms (1977) of data from Breder and Rosen (1966) and Budker (1971) relating to modes of reproduction in elasmobranchs shows that of the 16 families of Squaliformes, 12 are entirely viviparous, 2 are oviparous and 2 are mixed, and of the 12 Raiiformes families 9 are viviparous and 3 are oviparous. Earlier workers subdivided viviparous species into those that are ovoviviparous and those that are truly viviparous. However, this distinction is artificial and untenable since no known elasmobranch depends completely on its mother for nutriment during gestation (Dodd 1983). In view of this it is more informative to subdivide viviparous species into those that develop a placenta and those that do not (Budker 1958). Some aplacentals are solely dependent on yolk, some are provided with eggs to eat during gestation (oophagy) or cannibalise their siblings, and yet others are provided with 'embryotroph' or 'uterine milk' from a specially modified uterine epithelium (Wourms 1977). Placental species are those which develop a connection between the yolk sac of the embryos and the uterus of the mother at some time during gestation, often towards the end. Cyclicity, often strict, is usually associated with viviparity whereas oviparous species often appear to have protracted and ill-defined breeding seasons, though this is open to question.

Wourms (1977) and Dodd (1983) have recently reviewed what is known of reproduction in a range of species, oviparous, viviparous aplacental and viviparous placental. Here we shall take a single species representative of each mode and use it to illustrate the main structural and cyclical features associated with each.

Oviparity

The dogfish, *S. canicula* is an oviparous species which lays shelled eggs (mermaid's purses; Figure 12.3 inset) in pairs over what is usually considered to be an extended part of the year, though there are reasons for questioning this. Sumpter and Dodd (1979) have shown that in female dogfish, gonad size, amount of gonadotrophic hormone in the pituitary, and plasma sex steroids all follow a marked annual cycle and it is possible that breeding seasons in individual fish are shorter and more precise than is generally realised. The impression of an extended season may come from sampling populations that change due to migration. Furthermore, at least some of the records of gravidity on which cycle lengths are based are misleading because trawling stress can cause ovulation. 472 dogfish were taken during four trawling sessions in the same area (Dodd and Duggan 1982). In one group of fish examined immediately after trawling, 5.5 per cent of the fish contained fully-formed eggs presumably present before trawling started, while 13.8 per cent had purses in early stages of formation, which were believed to be products of trawling stress. In the three groups examined 20 hours after trawling the percentage of purses at all stages of formation, and including fully-formed purses, averaged 39.8. If it may be assumed that 5.5 per cent of all fish were gravid pre-trawl then about 34 per cent of the 20-hour fish were stress ovulated. If this interpretation is accepted, then, since all the data used to assess fecundity in the dogfish have been obtained from fish examined an appreciable time after trawling, they will clearly give a spuriously high estimate. Harris (1952), on the basis of such data, suggested that mature female dogfish must lay at least 10 eggs per month and that since the egg-laying season appears to last for at least ten months then the annual production must be at least 90 eggs; this is certainly too high an estimate.

There are several descriptions in the literature of copulatory behaviour in oviparous elasmobranchs based on aquarium observations (Wourms 1977; Dodd 1983). The only description based on observations made in the sea are those of A.C. Brooks reported by Dodd and Sumpter (1983) for *S. canicula*, as follows: 'the female lay straight and rigid but slightly tilted to the right. The male was coiled tightly around her pelvic region with his right flank in contact with the female's body and his ventral surface facing backwards. The left clasper lay across the right one, was curved through 90° and inserted into the female's cloaca. The right clasper was straight and occupied its normal position. The process is obviously protracted and the mating pair appeared oblivious to what was happening in their immediate vicinity. They were being harried by a group of about eight males which were swimming in tight circles around the copulating pair. One of these was seen to tug violently at the female's tail. It then moved round to face the female's head and carried out a similar assault whilst gripping her snout in its jaws. The female's only reaction was to close her eyes momentarily. After 20 minutes the pair were still motionless but had turned round, or been turned round, through 180°. One of the supernumerary males was lying on the bottom in contact with the female's head. Whilst still under observation, the copulating male, which had its eyes closed throughout the mating procedure, opened them and twitched its body slightly and several seconds later the female shot out of the loop formed by his body and swam off at high speed leaving the male writhing around, upside down, on the sea bed. The male was found to have the left clasper still bent and considerably frayed. These events took place in daylight at a depth of 50 feet in May.'

Aplacental Viviparity

One of the best documented accounts of reproduction in an aplacental viviparous species is that for *Squalus acanthias*, the spiny dogfish or 'spurdog' (Ford 1921; Te Winkel 1943; Templeman 1944; Hisaw and Albert 1947). Gestation lasts for 22 months and apart from the mating congregation, which in British waters takes place between November and January, the sexes are segregated. Hisaw and Albert (1947) have described the life history of the vast population of *S. acanthias* found off the eastern seaboard of North America. It migrates northwards in the spring and southwards in the autumn; the fish studied by Hisaw and Albert were abundant off Cape Cod in May and October-November. In May, all the females were pregnant, some in the early stages having recently ovulated and others carrying embryos of 12-20 cm in length, reflecting the 22-month gestation period. In the October-November fish there were again two groups, one of which contained females that may have given birth and others that had embryos that were 23-29 cm long, Hisaw and Albert's findings may be summarised as follows: ovulation probably occurs in February or March and the indications are that parturition occurs 22 months later, in late autumn, somewhere south of Wood's Hole, when the young are 25-30 cm long. Gestation is accompanied by strikingly long migrations. So far as the nutrition of the embryos is concerned, they certainly obtain water from the mother, but whether or not they obtain organic and inorganic nutrients is an open question, though the eggs contain more of both than do the fully developed young at birth. Corpora lutea, derived from ovulated follicles, are identifiable in the ovaries for most of the gestation period, but they undergo a progressive decrease in size accompanied by histological degeneration and there is no evidence that they serve any function.

Placental Viviparity

Placental viviparity is rare, being restricted to three families, Triakidae, Carcharhinidae and Sphyrnidae (Teshima 1981). The placenta is always of the yolk sac variety and the time at which it develops during gestation is variable. In some cases it is not developed until the middle of gestation and even at term embryos still contain some yolk in the intestine (*Mustelus griseus*; Teshima 1981). In yet others placentation commences soon after the start of embryonic development and the amount of yolk involved is negligible (*Scoliodon* spp.). In such species there is no egg membrane (Teshima 1981).

In *Carcharhinus dussumieri* (Teshima and Mizue 1972) only two oocytes are ovulated. A membrane, secreted by the nidamentary gland, invests each egg and continues to be secreted until it reaches about 40 cm in length. It is stored in a special chamber lying between oviduct and uterus. As development proceeds, the membrane emerges through a small hole to accommodate the growing embryo. When the latter reaches a length of 50 mm the base of the yolk sac comes in contact with the uterine wall, but there is still a large amount of yolk. This is used up by the time the embryo reaches a length of 150 mm and the placenta is established, the two being connected via the umbilical stalk. The latter contains an umbilical artery which connects with the dorsal aorta of the embryo, an umbilical vein which enters the hepatic vein and a vitelline duct which is connected with the intestine of the embryo. Parturition occurs when the embryos reach 370-380 mm, after a gestation period of unknown duration. Unusually for a viviparous species reproduction is said to be continuous throughout the year, though parturition appears to be more common in July and August than in other months.

Reproduction in Female Elasmobranchs

The structures associated with reproduction in female elasmobranchs are the ovaries, the reproductive ducts and the secondary sexual characters. Behaviour patterns associated with selecting mates, copulation, oviposition and parturition are also important.

The Ovaries

In all cases the ovaries originate as paired symmetrical structures, but in most species only the right ovary becomes mature (e.g. *S. canicula*). In a few cases the left ovary develops and in yet others both become functional (e.g. *Squalus acanthias* and oviparous skates). The mature ovary consists of a sparse stroma carrying blood vessels and nerves and a variable number of oocytes; oogonia, as we have seen, are absent. The appearance of the ovary is very variable between species and at different times in the reproductive cycle within species. But such differences are superficial and due largely to the presence or absence of yolky eggs which vary considerably in size and number; they may be as few as two, ranging 1-100 mm in diameter when ready for ovulation (Wourms 1977). Corpora atretica, which are yolky follicles that have failed to ovulate and are being resorbed, may be present in the mature ovary and also corpora lutea which are post-ovulatory follicles in various stages of reconstruction or atresia. A unique feature of the elasmobranch ovary is the fact that it is partially embedded in a lymphomyeloid structure called the epigonal organ; in the nurse shark *Ginglymostoma cirratum* this is said to average 0.60 per cent of the body weight and to produce granulocytes and lymphocytes (Fänge and Mattisson 1981)

The above description applies in general to most elasmobranch species so far described but, as we have seen, hardly anything is known of the large pelagic species. An exception is the basking shark *Cetorhinus maximus* whose reproductive organs have been described in detail by Matthews (1950). *C. maximus* is exceeded in size among fish only by *Rhineodon*, the whale shark; it may reach a length of 29 feet and a weight of 4 tons and its reproductive organs are built to scale! It is believed by Matthews (1950), on good evidence, to be viviparous, though a pregnant female has never been described, presumably because gestation occurs during an offshore migration. Only the right ovary persists. It is about 50 cm in length, elongated and oval in shape and fused with the anterior region of the epigonal organ. It contains little stroma and consists mainly of follicles of various sizes, corpora atretica, corpora lutea and blood vessels. The entire structure, which is probably unique is ramified by interconnected ducts which fuse to form passages of increasing diameter which finally converge on a 'pocket' opening into the body cavity on the right side of the ovary; the finest channels are ciliated. Matthews (1950) estimates that the ovary contains at least 6×10^6 oocytes, 0.5 mm or more in diameter and he believes that the yolky oocytes are ovulated at a diameter not greater than about 5 mm, since they must negotiate the fine ciliated passages in the ovary *en route* to the pocket through which they reach the body cavity and oviducts. Matthews gives a general description of follicle structure and development at optical microscope level and it is clear that they are basically similar to those of other vertebrates with yolky oocytes. In addition, Matthews describes two types of 'corpora lutea' in the basking shark ovary, and he supposes that the larger ones are post-ovulatory and the smaller ones corpora atretica in the sense used above. However, both from his histological descriptions and from the relative sizes of the two structures it is here suggested the reverse is true. The large numbers of corpora lutea remained unexplained by Matthews (1950), but in a later paper (Matthews 1955) he suggested that the ovulated eggs entered the oviducts during gestation and served to supplement the nutriment provided by the uterine trophonemata; this seems a likely explanation since such oophagy is practised by a number of aplacental viviparous species.

The basking shark ovary is unique amongst those that have so far been described and it is tempting to suggest that its structure is correlated with viviparity, but this may not be the case since the ovaries of viviparous species, both aplacental and placental, are much more similar to those of oviparous elasmobranchs such as the dogfish, *S. canicula* (Figure 12.1 a, b), which has been described by Metten (1939) and Dodd (1977). The appearance of the mature dogfish ovary, like that of other elasmobranchs, changes markedly throughout the year; in winter and early spring it contains a variable number of yolky oocytes of different sizes, previtellogenic oocytes, corpora atretica and corpora lutea. During the summer months (May-September) large yolky follicles that have failed to be ovulated become atretic and a crop of small growing oocytes can be identified from which the next year's eggs will be recruited. These grow throughout late summer and autumn at an equal rate with the shrinkage of the corpora atretica, but growth is not uniform; it results in a hierarchy of pairs of yolky follicles, about 30 in number, ranging in weight between 0.05 and 2.8 g (Figure 12.1b).

Follicular structure is close to that found in other vertebrates. The walls of the vitellogenic follicles consist of an outer layer of peritoneal epithelium, underlain by theca externa, theca interna, basement membrane and granulosa. The rich blood supply to the follicles lies between theca interna and granulosa and villi from the latter interdigitate with similar processes from the oocyte to form a zona radiata. The granulosa is multilayered in immature follicles, but becomes a single layer when vitellogenesis starts.

The only fine-structural study of the elasmobranch ovary is that of Dodd and Dodd (1980). The cells of the theca externa are cuboidal or flattened, and contain appreciable amounts of smooth endoplasmic reticulum, vesicles and mitochondria with tubular cristae. The theca interna is multilayered and contains large numbers of collagenocytes in a meshwork of collagen fibres. As the follicles grow, the thecal cells become dispersed in the collagenous matrix but may retain contact with each other by cellular processes. The interna cells also appear active, containing large amounts of rough endoplasmic reticulum and numerous mitochondria. They also contain intracellular actin-like fibrils, which by analogy with other vertebrates may be contractile and associated with ovulation, though nothing is known of this process in any elasmobranch. The granulosa cells have a fine structure reminiscent of high secretory activity with abundant rough endoplasmic reticulum, many mitochondria and numerous vesicles

Figure 12.1: Ovaries of *S. canicula*. (a) Ventral aspect of a mature ovary from a fish that is ovulating. Note vitellogenic oocyte (VO), corpus atreticum (CA), corpus luteum (post-ovulatory follicle; CL), epigonal organ (EO). (b) Mature ovary, vitellogenic oocytes (VO) dissected out to show hierarchy of sizes. The central region consists of stroma in which are embedded previtellogenic follicles of various sizes and which is surrounded by epigonal tissue. (c) Ventral aspect of an ovary of a mature female from which the thyroid gland has been removed. Note absence of vitellogenic oocytes. The epigonal organ is relatively more prominent than in (a) and (b). (The slips of paper do not refer to magnification.)

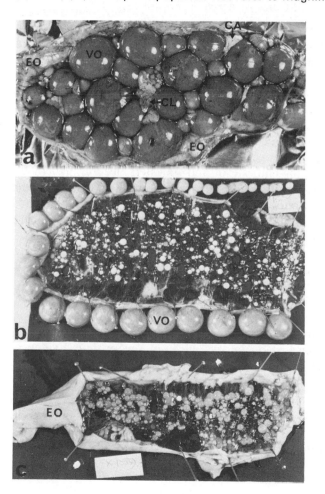

and granular inclusions of a range of different sizes. The granulosa is separated from the theca interna by a rich network of blood vessels and a prominent granular basement membrane which appears to contain fine striations. The granulosa cells are separated from each other by prominent intercellular spaces in which lie electron-dense inclusions believed to represent electron-lucent yolk precursor, which has passed from the blood vessels through the basement membrane to become electron-opaque in the medium contained in the intercellular spaces. This material reaches the zona radiata and is incorporated into the oocyte by pinocytosis.

Corpora Lutea and Corpora Atretica (Figure 12.1a)

The mature ovary, as we have seen, may also contain bright yellow bodies called 'corpora atretica' and also post-ovulatory structures called 'corpora lutea'. The former develop when yolk-containing oocytes undergo resorption. Chieffi (1962) has described the genesis of both types of structure in *Torpedo marmorata* and *T. ocellata*. In the formation of corpora atretica, cells of the granulosa proliferate to form well-vascularised villi, supported by thecal cells. These invade the oocyte, grow and anastomose and increase in vascularity. They absorb, digest and remove the yolk and become glandular, giving a positive reaction for cholesterol and its esters, leading Chieffi to suppose that they are functional endocrine structures homologous with the corpora lutea of mammals. However, Hisaw and Hisaw (1959) studied atresia in five species of elasmobranchs and concluded that it was merely a device for the removal of yolk from oocytes that failed to be ovulated. They also believed that any cellular activity seen in post-ovulatory structures was associated with removal of debris and that neither of these structures served any endocrine function. Chieffi (1962) however believes that 'true corpora lutea' are found in both oviparous and viviparous elasmobranchs. They are said to arise by atresia of preovulatory follicles in *Torpedo* and from post-ovulatory follicles in *Scyliorhinus*. Evidence for a functional significance is entirely indirect and comes from supposed correlations between their putative secretory activity and changes in reproductive structures, especially during gestation in viviparous species. Similarly, evidence of secretory activity is also indirect and based on the histochemical identification of the important steroidogenic enzyme 3β-hydroxysteroid dehydrogenase (3β-HSD). On the basis of such evidence Chieffi (1961) has suggested that in viviparous species the corpora atretica have an endocrine function whilst in oviparous species this function is sub-served by the post-ovulatory corpora lutea, but more recent evidence does not support such a generalisation (Lance and Callard 1969) and the question of the endocrine status of these structures remains open.

Vitellogenesis

It is now well established that most of the yolk in the eggs of verte-brates is synthesised by the liver under the stimulation of the female sex hormones. During vitellogenesis yolk precursor first appears in the plasma as a calcium-binding lipophosphoprotein called vitellogenin, which is incorporated into the oocytes, possibly under the influence of a gonadotrophic hormone. Early work on elasmobranchs appeared to show that they were exceptions to the general rule (Urist and Schjeide 1961) but the work of Craik (1978a, b, c, d) on *S. canicula*, recently reviewed by Dodd (1983), has shown that in fact they conform closely to the situation now well established in other vertebrates. A yolk granule-like phosphoprotein, which has not as yet been isolated and characterised, occurs in vitellogenic female dogfish. It is present in relatively low amounts though this is probably correlated with the protracted egg-laying season of the dogfish, during which it ovulates only two eggs at intervals of 1-3 weeks. Similarly, the annual cycle of plasma levels of phosphoprotein is much less marked than in verte-brates with a more restricted breeding season, though the levels between September and March, the peak egg-laying period, are significantly higher than between March and August (Craik 1978c). Craik (1978d) has also shown that vitellogenin synthesis in the dogfish, as in all other vertebrates, appears to be stimulated by oestradiol; intramuscular injections of oestradiol-17β (3 mg per kg) produced a 17-fold increase in levels of plasma phosphoproteins, whilst only increasing circulating oestradiol levels by a factor of two to three.

Steroidogenesis

Rather little is known about the sex steroids in elasmobranchs and even less about their functions. However, steroids have been identified in ovarian extracts and steroidogenic enzymes have been localised in various regions of the ovary, especially follicle wall cells. Ovarian tissues are capable of steroid synthesis *in vitro*. Extracts of *Squalus suckleyi* ovaries have been shown to contain oestradiol-17β (120 µg per kg) with traces of progesterone and oestrone (Wotiz, Botticelli, Hisaw and Bingler 1958; Wotiz, Botticelli, Hisaw and

Olsen 1960). Chieffi and Lupo di Prisco (1963) found progesterone, oestriol and oestradiol-17β in *Torpedo marmorata* ovarian extracts and oestradiol-17β and oestrone have been identified in *S. canicula* and *Squalus acanthias* (Simpson, Wright and Hunt 1963; Gottfried 1964). There have been a number of studies on the occurrence and distribution of steroidogenic enzymes of which the most detailed is that of Lance and Callard (1969) in *S. acanthias.* They demonstrated that glucose-6-phosphate dehydrogenase (G-6-PDH) was present in all ovarian tissues tested, 3β-hydroxysteroid dehydrogenase (3β-HSD) was found in the granulosa and theca externa of developing follicles, though not in the theca interna, and it was also found in post-ovulatory follicles, though not in corpora atretica. The status of these structures in steroid synthesis is confused. Chieffi (1961) has reported that in *Torpedo marmorata* and *T. ocellata* only the corpora atretica are steroidogenic, whereas in *S. stellaris* and several oviparous skates it is the post-ovulatory follicles that are secretory. Clearly, in the absence of direct evidence it is impossible to decide whether or not they serve any endocrine function. They may have roles in reproduction but the evidence is difficult to evaluate and there is a need for more work.

The few *in vitro* studies on steroidogenesis by the elasmobranch ovary indicate that it produces oestradiol-17β, testosterone and small amounts of progesterone. Our own recent studies on isolated ovarian tissues of *S. canicula* (Figure 12.2) reveal that the biosynthetic capacity of follicle wall cells is related to follicle size. Thus, small vitellogenic follicles produce large quantities of oestradiol-17β (E2) but considerably less testosterone (T), the ratio E2:T being approximately 10:1; the overall rate of steroidogenesis increases with increasing follicular maturity up to follicle weights of about 600 mg, when production rates may reach 1300 pg oestradiol-17β and 400 pg testosterone per mg follicle wall per hour. In follicles of 1-1.5 g, steroidogenesis, especially aromatisation, is less vigorous, with a consequent reversal of the E2:T ratio. Thereafter, steroid production diminishes sharply in follicles approaching ovulatable size (about 2.5 g), when output of both oestradiol-17β and testosterone is less than 5 per cent of that achieved at peak production. Even lower rates of production of the two steroids are found in post-ovulatory and atretic follicles. Analysis of the ovarian incubates for progesterone shows that little is produced by either the follicles or the post-ovulatory follicles.

Both oestradiol-17β and testosterone are found in the plasma of *S.*

Figure 12.2: (a) Steroid content (testosterone and oestradiol-17β) of ovarian follicular walls in pg per mg tissue (wet weight). FW, wet weight of freshly dissected tissue in mg; POF, post-ovulatory follicle. (b) *In vitro* production of testosterone, oestradiol-17β and progesterone by ovarian follicle walls. FOLLS, wet weight of freshly dissected follicular tissue in g; POF, post-ovulatory follicle; AF, atretic follicle. Note that ordinates are plotted on a logarithmic scale.

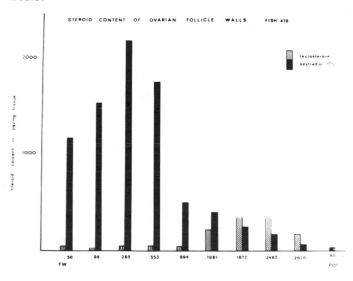

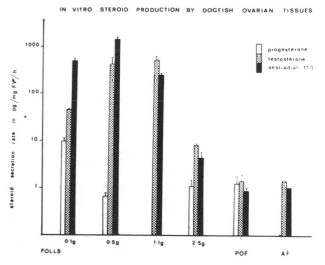

canicula and display a clear, synchronous annual cycle rising with the recrudescence of the ovary to reach high levels during the active months of the extended egg-laying season. Plasma testosterone levels are surprisingly high thoughout the year in females, at approximtely 20 per cent those of oestradiol-17β. Whilst oestradiol in the female is implicated in vitellogenesis and the development and maintenance of secondary sexual characteristics such as the shell glands and oviducts (Dodd and Goddard 1961), the role of testosterone in the female is less clear. It seems likely that it fulfils a specific role in reproduction, perhaps in the control of reproductive behaviour. Areas of the brain of the dogfish are known to bind androgens (Jenkins 1978) and Callard, Petro and Ryan (1978a, b) have shown that aromatases and other androgen-converting enzymes are present in the elasmobranch brain; thus it is possible that testosterone functions at the level of the central nervous system.

The physiological significance of the quantitative differences in steroid output by follicles of different sizes is unknown. Differential steroid production may play a part in any one of several processes, for example the establishment and maintenance of the follicular hierarchy itself, or in determining the sequence of follicles selected for ovulation. Nor do we understand how ovarian steroidogenesis is controlled by the pituitary; we have been unable to stimulate or otherwise modify *in vitro* steroid production by ovarian tissue with pituitary preparations, although preliminary results show clearly that removal of the ventral lobe of the pituitary of donor fish three weeks before experimentation markedly reduces the rate of ovarian steroidogenesis *in vitro*, and reduces aromatisation.

Reproductive Ducts (Figure 12.3)

The oviducts of elasmobranchs are derived from the paired pronephric or Müllerian ducts of the embryo. Both usually develop (but there are exceptions) and become differentiated into a number of regions serving different functions. The histology and degree of development of the regions depend largely on the state of maturity of the fish and the mode of reproduction, whether oviparous or viviparous. The anterior regions of the ducts are usually short and unspecialised; they open into the body cavity by the osteum, a slit-like aperture lying in the falciform ligament. Elasmobranch ovaries are gymnoarian, that is, the eggs are shed into the body cavity and carried to the osteum by cilia. The first specialised region in the oviduct is the oviducal or nidamentary gland, best developed in oviparous species.

Figure 12.3: Internal reproductive organs of a mature female dogfish, right oviduct dissected to show internal structure. A, albumen gland; L, liver; LOD, lower oviduct (note muscular walls); MOD, middle oviduct; O, oesophagus; OS, osteum abdominale (through which the oviducts open into the body cavity); OV, ovary; P, left pectoral fin; S, shell gland; T, 'tendril' material secreted by the shell gland; UOD, upper oviduct; V, vagina. *Inset:* Mermaid's purses. a, case with no egg; b and c, purses containing eggs and albumen. Note that the purses b and c differ in form. Each female lays purses that are specific and constant in form.

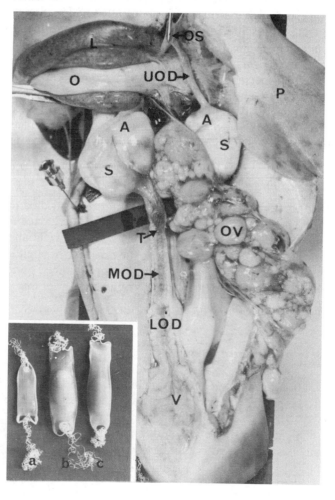

It has been described in *S. canicula* by Metten (1939) in which it is large and ovoid and consists of three distinct regions, an anterior albumen-secreting zone, a short mucous-secreting zone and a large posterior region by which the shell or mermaid's purse is secreted, and which acts as a receptaculum seminis. Sperm survival of up to 5-6 weeks in this structure has been reported by Clark (1922) in the skate *Raia brachyura* and in our own work on *S. canicula* we have encountered striking sperm longevity: a number of females have laid fertile eggs up to 15 months after contact with a male and one specimen caught on 5 April 1979 laid a pair of fertile eggs on 23 April 1981, more than two years after any possible insemination.

The oviducal gland in viviparous species is usually smaller, though all three regions are represented and functional. The shell is more like a membrane in such species, though it usually persists throughout development and may form part of the placenta (Teshima and Mizue 1972). In viviparous species the lower oviduct becomes a 'uterus' during gestation and its inner lining differentiates into complicated glandular folds and villi called trophonemata which produce embryotroph on which the developing embryos feed. In oviparous species the lowest part of the oviduct is usually a muscular region and in immature elasmobranchs it is closed by a membraneous hymen.

Secondary Sexual Characters

External secondary sexual characters in the female are restricted to larger body size and a more distended body cavity. The cloaca is larger and more richly endowed with mucous-secreting cells and sense cells. Internal sex differences comprise the presence of vitellogenin in the plasma throughout most of the year, a pronounced cycle in liver:body weight ratio and extensive areas of ciliation over many of the viscera.

Reproduction in Male Elasmobranchs

Unlike the situation in female elasmobranchs, the gonads in males seem always to be paired. They consist of two testes, usually large relative to body size, suspended from the dorsal body wall by mesorchia which carry blood vessels and nerves. Like the ovaries, each is intimately associated with an epigonal organ.

The Testes (Figure 12.4)

Testis structure is unusual in two respects: the unit structure is the ampulla or follicle rather than the tubule, and the ampullae are zonate, being arranged in concentric bands which radiate outwards from an ampullogenic zone lying along the lateral or dorsolateral margin of the testis. Furthermore, each ampulla contains germ cells at an identical stage of spermatogenesis and each zone consists of ampullae at the same stage (Figure 12.4). Mellinger (1965) has

Figure 12.4: (a) Diagrammatic representation of zonate testis and spermatogenesis in *S. canicula* (transverse section). I, aggregation of gonia and Sertoli cells; II, ampullogenic zone; III, spermatogonia; IV, primary and secondary spermatocytes; V, spermatids and spermateliosis; VI, spermatozoa; VII, release of spermatozoa; VIII, ampullae containing only Sertoli cells (From Dodd 1983) (b) T.S. normal testis showing spermatogonial ampullae. Note that Sertoli cells are perilumenar. Some of the spermatogonia are undergoing mitosis; note that this is synchronous within spermatocysts. (c) T.S. two spermatozoa ampullae. Sertoli cells with their prominent nuclei lie peripherally, each containing 64 spermatozoa.

a

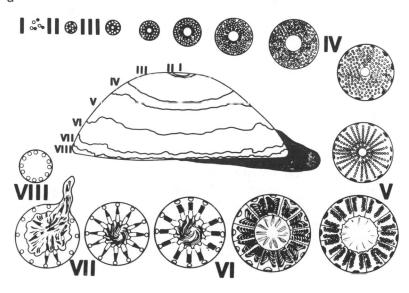

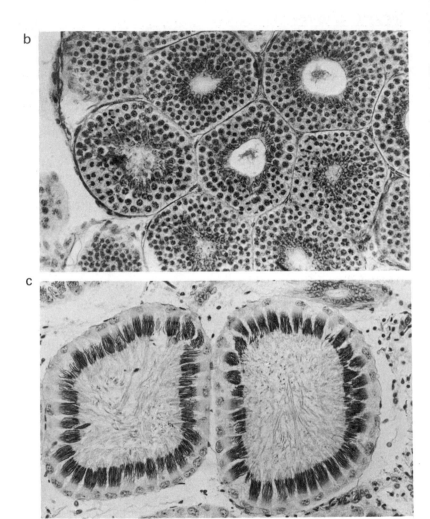

reviewed the early literature on the testis in elasmobranchs and described in detail the structure of the *S. canicula* testis. Stanley (1966) has done the same for *S. canicula* and *Torpedo marmorata*, demonstrating for the first time the special relationship between Sertoli cells and germ cells, and correcting some earlier misconceptions. The ampullogenic zone consists of gonia, Sertoli cells and fibroblasts (Figure 12.4a). The gonia are easily separable from the other cells by their large nuclei and prominent nucleoli. At a number of loci in the ampullogenic zone the two types of cells come together in groups around which an investment of fibroblasts forms. At first the nests are solid but a central lumen soon appears and both types of cells undergo repeated mitotic divisions, the Sertoli cells arranging themselves in a ring around the lumen, the germ cells lying peripherally (Figure 12.4b). In *S. canicula*, the divisions produce approximately 500 Sertoli cells and 50 spermatogonia and Stanley has shown by phase contrast microscopy that each Sertoli cell engulfs a single spermatogonium forming a spermatocyst. All subsequent divisions of the germ cells (4 mitotic, 2 meiotic) take place within the spermatocyst: thus each ampulla produces about 32 000 spermatozoa. By the end of the spermatogonial stage in *S. canicula* the follicles have increased markedly in diameter from about 150 μm to 225 μm and the Sertoli cells have migrated to lie peripherally. Ampullae containing primary spermatocytes are relatively rare and this is believed to reflect the short duration of this stage. The spermatids come to lie around the margins of the spermatocyst (Sertoli cell) wall and during spermateliosis, which Stanley (1971a, b) has described in detail in *Squalus suckleyi*, the developing sperm tails project into the lumen of the ampulla. When sperm development is complete, the heads of the 64 spermatozoa become embedded in the residual cytoplasm of each Sertoli cell (Figure 12.4c). Until the sperm are ready to be shed the ampullae are closed, though each is connected to a ramifying system of efferent ducts from the earliest stages. The fully mature follicles acquire an opening into the duct system, the latter opening into the vasa efferentia and vas deferens. Once the sperm leave the follicles, the latter contain only Sertoli cell nuclei and debris and they degenerate and are resorbed. Dobson (1974) has analysed testis weight and ampulla constitution in *S. canicula* and shown that the changes are cyclical even though mature sperm can be expressed from the urinogenital papilla at all times of the year. In *Squalus acanthias* cyclicity is more pronounced and Simpson and Wardle (1967) have shown that all primary spermatocytes are transformed into spermatozoa within a 12-month period and maximum sperm development occurs between November and January, the time at which insemination is known to occur. A zone of degeneration that first appeared in May of each year, was identified; the significance of this is discussed below. A similar annual testis cycle has been described in *Mustelus griseus* and *M. manazo*, seasonal breeding smooth dogfishes of Japan. Insemination occurs between June and August and the cycle is geared to maximum production of sperm at this period; spermatogenesis is reinitiated immediately after mating (Teshima 1981).

Steroidogenesis

In most vertebrates it is accepted that the main source of steroids in the testis is the interstitial tissue (Leydig cells) though lobule boundary cells and Sertoli cells have also been implicated. In elasmobranchs the presence of histologically recognisable interstitial tissue has been questioned (see Dodd 1983 for a review) though there is no doubt that steroids are synthesised by the testis in these animals. Chieffi and Lupo di Prisco (1961) have shown that extracts of the testis of *S. canicula* contain testosterone, androstenedione, progesterone and oestradiol. Simpson, Wright and Hunt (1964) demonstrated that testis tissue from *S. acanthias* can synthesise testosterone from ^{14}C-progesterone and Kime (1978) found that *in vitro* incubation of testicular tissue of *S. canicula* with ^{14}C-progesterone or pregnenolone yields androstenedione and testosterone. The identity of the steroid-secreting tissue is still in doubt though Chieffi, Della Corte and Botte (1961) and Della Corte, Botte and Chieffi (1961) have produced convincing evidence that in *Torpedo marmorata* and *S. stellaris* at least, the testis contains typical interstitial cells giving a positive reaction for 3β-HSD. Dodd (1972) suggested that the Sertoli cells might be steroidogenic and have a role in the control of spermatogenesis. This possibility is supported by the demonstration that in *S. canicula* and *Squalus acanthias* they contain 3β-HSD (Collenot and Ozon 1964; Simpson and Wardle 1967).

Reproductive Ducts

The male ducts consist of vasa efferentia, derived from mesonephric tubules of the embryo and the vas deferens which originates as the mesonephric kidney duct or Wolffian duct, the anterior end of which forms a coiled epididymis whilst the posterior region frequently has diverticulae or seminal vesicles. In the basking shark, the vas deferens is modified to form a large ampulla with a complex internal structure and a capacity of 'several gallons' in which spermatozoa are packaged into spermatophores (Matthews 1950).

Secondary Sexual Characters

Male elasmobranchs at maturity are smaller than females and their body contours are more slender. Sexual maturity is reached earlier, they have a shorter life span and are more active and aggressive. Some skates have spines on their wings which are claw-like and retractile (Mellinger 1966a). But the most striking external secondary sexual characters in the male are the claspers, structures involved in copulation which, in some skates, assume grotesque proportions. They are derived from the margins of the pelvic fins, have a jointed cartilaginous skeleton, are usually armed with large placoid scales and contain a central canal which communicates with a muscular sac or siphon. One, or both, claspers are used to distend the female's cloaca during copulation and the siphon is believed to aspirate sea water over the urogenital papilla of the male and so flush sperm into the oviducts.

Endocrine Regulators of Reproduction

The Pituitary Gland

Structure (Figure 12.5)   Only a few elasmobranch species have been studied in the context of the pituitary gland and in only two, *S. canicula* and *Mustelus canis* have the effects of pituitary removal been investigated. The gland shows a good deal of superficial morphological variation in different species (Norris 1941) but it is probably reasonable to generalise that it always shows a degree of subdivision into more or less separate lobes and that a markedly segregated ventral lobe is always present (Knowles, Vollrath and Meurling 1975; Dobson and Dodd 1977a, b; Dodd and Sumpter 1983). The adenohypophysis consists of a pars intermedia, which is closeley associated with the neurohypophysis to form a neuro-intermediate lobe (NIL), a tongue-like region comprising an anterior rostral lobe (RL) and a posterior median lobe (ML) and, connected to the latter by an interhypophysial stalk, a ventral lobe (VL).

In sharks and dogfish the VL is deeply embedded in the basal cartilage of the skull, whereas in skates and rays it is usually strap-like, lies intracranially in a shallow groove and has a more substantial connection with the ML. All the lobes have been said at one time or another to be connected with reproductive control but there is general agreement that the VL is much the most important (Dodd, Evennett and Goddard 1960; Mellinger 1963, 1965, 1966b; Holmes and Ball 1974). Circumstantial evidence for a function in reproduction comes from tinctorial and semi-histochemical identification of putative gonadotrophic cells. Te Winkel (1969) identified cells in the VL of *Mustelus canis* which were periodic acid-Schiff-positive and

Figure 12.5: Topography and vascularisation of the dogfish pituitary (from Dodd 1975). (a) Photograph of the pituitary complex *in situ*; the blood vessels have been injected with Indian ink. Note that the ventral lobe (V), is supplied with blood from the anastomosis (ca), of the carotid vessels (c). li, lobi inferiores of the hypothalamus; m, median lobe; n, neurointermediate lobe; oc, optic chiasma; on, optic nerve; r, rostral lobe; s, saccus vasculosus. (Bottom is anterior): (b) Diagrams of the pituitary lobes in sagittal section (upper) and in ventral aspect (lower). Li, lobi inferiores; Me, median eminence; ML, median lobe; NIL, neurointermediate lobe; Oc, optic chiasma; RL, rostral lobe; S, saccus vasculosus; St, stalk connecting ventral and median lobes; VL, ventrial lobe; III, third ventricle.

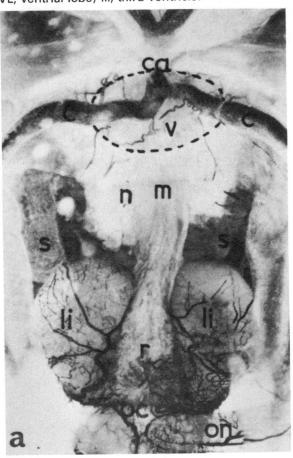

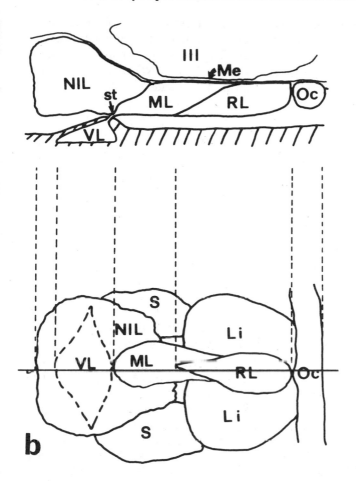

aldehyde fuchsin positive and suggested that these were gonadotrophs. Knowles *et al.* (1975), in a fine-structural study of the
pituitary of *S. canicula,* have identified putative gonadotrophs based
on granule size; Mellinger and Dubois (1973) have shown that cells
are present in the VL of *Torpedo marmorata* that react with a
fluorescein-labelled antibody to ovine luteinising hormone (LH);
and we have obtained similar results in *S. canicula* using a fluorescent
antibody to quail LH. However, identifications based on granule size
and heterologous antibodies are suspect. So far as the other lobes are
concerned, Knowles *et al.* (1975) have described cells in the ML of *S.
canicula* which, based on granule size, they tentatively identified as

gonadotrophs, and Chieffi (1967) has reviewed histological work on changes in lobes of the pituitary other than the VL possibly related to reproductive events in both oviparous and viviparous species, including the appearance of giant cells in the NIL of the viviparous *T. marmorata* during pregestation. These studies indicate that other lobes of the pituitary may also be involved in reproductive control.

This indirect evidence that the VL is the main gonadotrophic lobe has been supplemented by surgical investigations. As we have seen, the pituitary is markedly subdivided into regions and this enables the lobes to be removed separately; the VL, because of its situation, can be entirely removed without entering the cranial cavity (Dobson and Dodd 1977a). When this is done in the male dogfish at temperatures in excess of about 13°C, a highly characteristic band of degeneration is produced in the testis, precisely localised in the zone containing late spermatogonia. The absence of the VL appears to affect spermatogonia about to undergo the last mitotic division (which converts them into primary spermatocytes). No other stages in spermatogenesis are affected and ablation of the other pituitary lobes, either separately or together, is ineffective. The importance of the VL in stimulating mitotic activity in the testis is reflected also in the reduction in uptake of ^{3}H-thymidine by the testis which follows removal of this lobe (Dobson and Dodd 1977a).

Ventral lobectomy, but not removal of the other lobes, in the female dogfish results in atresia of all the yolky follicles and prevents subsequent growth of the ovary (Dodd 1972). Thus in elasmobranchs, as in other vertebrates, the pituitary is essential for gonad maintenance in both sexes, the gonadotrophic effect in these fish being localised mainly or entirely in the VL. As we have said, there is indirect evidence that the other pituitary lobes may be involved in the control of reproduction, but surgical removal of them has so far revealed no effect on gonad maintenance.

Hormones (1)   *Gonadotrophin (s)* (GTH).   The subdivision of the elasmobranch pituitary gland into discrete lobes has facilitated the investigation of its role in reproduction in a number of ways. Removal of each lobe separately has shown that the gonadotrophic function is mainly localised in the ventral lobe (Dodd *et al.* 1960; Mellinger 1963). Furthermore, saline extract of the separate lobes have been tested for activity in a number of heterologous bioassays for gonadotrophin, including stimulation of uptake of ^{32}P by the testes of day-old chicks (Scanes, Dobson, Follett and Dodd 1972), or androgen production by isolated testicular cells of quail (QTCA; Sumpter *et al.* 1978a, b) or turtle (Lance and Callard 1978). Again, almost all of the biological activity has been found in the VL, with little activity in the other lobes. In the dogfish, *Scyliorhinus canicula* the VL contains 98.8 per cent of the total GTH activity, the NIL and combined RL and ML containing 0.9 and 0.3 per cent respectively, when measured in the QTCA (Sumpter *et al.* 1978b).

The amount of GTH activity found in the VL is not constant throughout the year, at least in females. When dogfish VL extracts, prepared from monthly samples, were assayed for GTH by bioassay, activity was seen to rise gradually from a seasonal low of less than 1.25 µg equivalents per VL (the potency here is expressed as 'mass equivalents' of a reference gonadotrophin, National Institute of Health ovine LH) in July to peak during mid egg-laying season (40-80 µg eq/VL February-April; Sumpter and Dodd 1979).

Whilst it is possible to quantify pituitary VL gonadotrophic hormone content (in ovine LH equivalents) by bioassay, our inability to measure peripheral plasma concentrations of GTH has hampered attempts to look more closely at pituitary control of gonadal function in elasmobranchs. Since current bioassays are not sufficiently sensitive, we have sought to develop a suitably sensitive and hormone-specific radioimmunoassay (RIA) for this purpose, but this has not yet been successful.

(2)   *Thyrotrophin* (TSH).   The TSHs of other vertebrates are, like the GTHs, glycoproteins. The extraction of TSH activity in the glycoprotein fraction from dogfish VLs suggests a similar relationship in the elasmobranch, although few data are available. There is some evidence that in the dogfish TSH exists as a moeity separate from GTH because as the potency of GTH increases during purification, that of TSH decreases in a manner suggestive of the removal of a similar yet separable molecule (Sumpter *et al.* 1978b). Attempts to establish an homologous *in vitro* bioassay for TSH have failed to produce consistent results.

Until we have reliable homologous bioassays for both GTH and TSH, it will not be possible to ascribe roles to the glycoproteins isolated from the VL, regardless of their activities in heterologous assays.

The Hypothalamus

The role of the hypothalamus in the control of pituitary hormone secretion and release is well established in most vertebrates. Control is mediated via neurosecretory hormones elaborated in hypothalamic neurons and brought either to the median eminence, a neurohaemal organ, for release into the pituitary portal vessels, or carried directly by axons to pituitary cells as in most teleosts and the NIL of elasmobranchs. The elasmobranch hypothalamus consists laterally of the walls of the third ventricle and ventrally of a thin sheet of tissue in which neurosecretory axons run and in which lies the median eminence, divisible into anterior and posterior regions. The former connects with the RL and the latter with the ML and NIL. So far as is known, the VL has no direct connection with either part of the median eminence, being vascularised by the internal carotid arteries (Mellinger 1960, 1961; Meurling 1960, 1967). Thus the relationship between the median eminence and the pituitary, especially its main gonadotrophic region, is atypical and the question as to if and how the VL activities are controlled by the hypothalamus remains open.

Four neurosecretory nuclei have been identified in the hypothalamus of a number of elasmobranchs (Mellinger 1964; Meurling 1967; Munro and Dodd 1983) though the destinations of their axons, other than those of the preoptic nucleus (PON) are largely unknown. Some of the latter end in the posterior region of the median eminence and Knowles *et al.* (1975) have identified aminergic neurons of unknown origin which directly innervate putative gonadotrophs in the ML. Thus if the ML does, in fact, subserve a gonadotrophic function a morphological basis exists for neurosecretory control. But there is no unequivocal evidence that releaser substances are produced by the elasmobranch hypothalamus although several workers have provided circumstantial evidence that they are. King and Millar (1979, 1980) (by radioimmunoassay using an antibody to synthetic mammalian GnRH) found cross-reaction to hypothalamic extracts of the dogfish *Poroderma africanum*, though Deery (1974) obtained negative results in similar studies on *S. canicula*. Deery and Jones (1974) found that extracts of median eminence in this species activated the adenyl cyclase system in all the pituitary lobes, though synthetic GnRH activated only the Vl enzyme. Jenkins and Dodd (1980) have shown that intravenous injection of the synthetic releaser stimulates a rise in circulating sex steroids in both sexes of *S. canicula* and increases the rate of oviposition. Similar results were obtained when neutralised acid extracts of dogfish hypothalamus were used and the latencies were similar. Thus it appears that the elasmobranch hypothalamus is involved in reproductive control as in other vertebrates, though its mode of action remains in doubt, especially in connection with the VL where evidence shows that releaser substances could reach this lobe only by a systemic route. This receives support from the recent findings of King and Millar (1980) that immunoreactive GnRH is present in unusually high concentrations (0.6 ng per ml) in the blood of the dogfish *Poroderma africanum*.

Recent studies of Jenkins, Joss and Dodd (1980) have provided additional evidence that the hypothalamus has a role in reproductive regulation in *S. canicula*. They have shown that the hypothalamus contains a high concentration of oestradiol-specific receptors, localised in the preoptic, habenular and tuberal neurosecretory nuclei and suggested that they may function, through feedback, to regulate circulatory gonadotrophin levels.

The Thyroid Gland

The elasmobranch thyroid, unlike that of teleosts, is discrete and encapsulated and easily removed surgically. As in all vertebrates, it is follicular and secretes two hormones: thyroxine (T_4) and triiodo-thyronine (T_3). The role of the thyroid in poikilotherms has proved particularly difficult to identify apart from two striking exceptions: amphibian metamorphosis and smoltification in salmon. In elasmo-branchs there is a good deal of circumstantial evidence that thyroid activity is correlated with reproduction (Dodd and Sumpter 1983) but direct evidence is lacking. In addition to various histological and biochemical correlations between thyroid activity and reproduction, Olivereau (1954), Matty (1960) and Lewis and Dodd (1974) have reported that the gland, as reflected by the thyrosomatic index (TSI), is at least twice as large in adult female elasmobranchs as in males. Indeed, Mellinger (1966a) states that this dimorphism is the most spectacular of all the secondary sexual characters. We have also found (unpubl. obs.) that the TSI reaches its highest value in females approaching puberty. These circumstantial indications of an important correlation between reproduction and the thyroid in the female have now been reinforced by direct evidence obtained after surgical removal of the gland. Lewis and Dodd (1974) have shown that the thyroid is essential for the annual gonadal recrudescence that takes place in *S. canicula* each autumn after a period of atresia during the summer months (Figure 12.1c), though the mode of action of the thyroid hormones has not yet been discovered. It seems likely that they either stimulate vitellogenin synthesis in the liver, or are implicated in yolk uptake by the follicles. It is not known whether the thyroid has a role in the reproductive physiology of the male.

Summary

The elasmobranchs have many characteristics to make them of particular interest to comparative reproductive physiologists. They are phylogenetically amongst the oldest of the jawed vertebrates; they have reproductive ducts similar in structure and ontogeny to those of amniotes, unlike the teleosts; internal fertilisation is universal and viviparity, including placental viviparity, is widespread; and they have a full complement of the endocrine organs associated with reproduction in vertebrates (Wourms 1977). Yet, as the above review shows, they have received surprisingly little attention. Even at the level of descriptive biology, histology and histochemistry only restricted aspects and few species have been studied. Experimental investigations have been virtually restricted to two squaliform species. This has led to generalisations that may prove to be unwarranted when more species are studied. Indeed, there is reason to believe that this may well be the case in a group of animals as old as this, in which many of the extant genera were already established before there were marine bony fish and in which the two main subgroups, squaliformes and raiiformes are themselves of Mesozoic origin.

The gonads, in the few species investigated, and with the exception of the basking shark ovary, are closely similar in structure and it seems reasonable to suppose that they are likely to be under trophic control by the ventral lobe of the pituitary, but nothing is known with certainty of the ways in which this control is regulated or whether the ventral lobe always lacks a hypothalamic portal blood supply. Similarly, we are ignorant of the means by which the hierarchy of follicle sizes is initiated and maintained in oviparous species and the way in which ovulation is controlled. Mating behaviour associated with insemination is sophisticated in the few species where it is known, but here again nothing is known of the agencies that control it.

Recent work on the thyroid, as we have seen, has shown unequivocally that in the female dogfish, its hormones are essential for gonadal growth (presumably vitellogenesis), though whether they act at the level of the liver or the ovarian follicles is not known.

Another major gap in our knowledge of elasmobranch reproductive physiology concerns environmental regulators. In contrast to the teleosts, in which a good deal is known about the environmental cues that control reproductive cycles, virtually nothing is known concerning elasmobranchs. They have a pineal gland, which has the ultrastructural characteristics of a light-sensitive organ. There is tentative evidence that removal of the pineal in the female dogfish results in an increase in the gonadotrophic potency of the ventral lobe, whilst injection of pineal extracts into pinealectomised fish reduces its gonadotrophin content to non-detectable levels. These findings may indicate that light is an environmental regulator, but much more work is necessary before this can be stated with any confidence. Our own work on the dogfish suggests that temperature is the main environmental trigger (Dobson and Dodd 1977b). Certainly, reproductive activity as shown by ovarian growth, ovulation and oviposition is maximal during winter when temperatures are low, but here again there is a case for believing that environmental cues might well be different for fish living in waters of relatively constant temperature when light might be a more appropriate cue.

This summary exposes only some of the many gaps in our knowledge of the reproductive physiology of elasmobranchs and reiterates the dangers of generalising until a much greater range of species has been investigated. Until this has been done we will not have a comprehensive picture of reproduction in these interesting and important animals.

REFERENCES

Ablett, R.F., Sinnhuber, R.O. and Selivonchick, D.P. (1981) 'The effect of bovine insulin on ¹⁴C-glucose and ³H-leucine incorporation in fed and fasted rainbow trout *(Salmo gairdneri).*' *General & Comparative Endocrinology, 44* : 418-27.

Ablett, R.F., Sinnhuber, R.O., Holmes, R.M. and Selivonchick, D.P. (1981) 'The effect of prolonged administration of bovine insulin in rainbow trout *(Salmo gairdneri).*' *General & Comparative Endocrinology, 43* : 211-17.

Abrahamsson, T. and Nilsson, S. (1976) 'Phenylethanolamine-N-methyl transferase (PNMT) activity and catecholamine content in chromaffin tissue and sympathetic neurons in the cod, *Gadus morhua.*' *Acta Physiologica Scandinavica, 96* : 94-9.

Abrahamsson, T., Holmgren, S., Nilsson, S. and Pettersson, K. (1979a) 'On the chromaffin system of the African lungfish, *Protopterus aethiopicus.*' *Acta Physiologica Scandinavica. 107* : 135-9.

Abrahamsson, T., Holmgren, S., Nilsson, S. and Pettersson, K. (1979b) 'Adrenergic and cholinergic effects on the heart, the lung and the spleen of the African lungfish, *Protopterus aethiopicus.*' *Acta Physiologica Scandinavica, 107* : 141-7.

Ahlqvist, R.P. (1948) 'A study of the adrenotropic receptors.' *American Journal of Physiology, 153* : 586-600.

Ahmad, M.M. and Matty, A.J. (1975a) 'Effect of insulin on the incorporation of ¹⁴C-leucine into goldfish muscle protein.' *Biologia (Lahore), 21* : 119-24.

Ahmad, M.M. and Matty, A.J. (1957b) 'Insulin secretion in response to high and low protein ingestion in rainbow trout *(Salmo gairdneri).*' *Pakistan Journal of Zoology, 7* : 1-6.

Akert, K. (1949) 'Der visueller Greifreflex.' *Helvetica Physiologica et Pharmacologica Acta, 7* : 112-34.

Akhtar, S., Butt, M.A. and Ahmad, M.M. (1979) 'Effect of insulin and glucose loading on the blood sugar level of the freshwater teleost, *Channa punctatus* (Bloch).' *Pakistan Journal of Zoology, 11* :179-81.

Allison, L.N. (1951) 'Delay of spawning in eastern brook trout by means of artificially prolonged light intervals.' *Progressive Fish Culturist, 13* : 111-16.

Alt, J.M., Stolte, H., Eisenbach, G.M. and Walvig, F. (1981) 'Renal electrolyte and fluid excretion in the Atlantic hagfish *Myxine glutinosa.*' *Journal of Experimental Biology, 91* : 323-30.

Aronson, L.R. (1981) 'The evolution of telencephalic function in lower vertebrates' in P. Laming (ed.) *Brain Mechanisms and Behaviour in Lower Vertebrates,* University Press, Cambridge, U.K., pp. 33-58.

Ask, J.A., Stene-Larsen, G. and Helle, K.B. (1980) 'Atrial beta-adrenoceptors in the trout.' *Journal of Comparative Physiology, 139B* : 109-16.

Babiker, M.M. and Rankin, J.C. (1978) 'Neurohypophysial hormonal control of kidney function in the European eel *(Anguilla anguilla L.)* adapted to sea-water or fresh water.' *Journal of Endocrinology, 76* : 347-58.

Babiker, M.M. and Rankin, J.C. (1979a) 'Factors regulating the functioning of the *in vitro* perfused aglomerular kidney of the anglerfish, *Lophius piscatorius L.*' *Comparative Biochemistry and Physiology, 62A* : 989-93.

Babiker, M.M. and Rankin, J.C. (1979b) 'Renal and vascular effects of neurohypophysial hormones in the African lungfish *Protopterus annectens*

(Owen).' *General & Comparative Endocrinology, 37* : 26-34.

Baerands, G.P. (1975) 'An evaluation of the conflict hypothesis as an explanatory principle for the evolution of displays' in G.P. Baerands, C. Beer & A. Manning (eds.), *Function & Evolution in Behaviour,* Clarendon Press, Oxford, pp. 225-73.

Baggerman, B. (1966) 'On the endocrine control of reproductive behaviour in the male three-spined stickleback *(Gasterosteus aculeatus L.).*' *Society for Experimental Biology Symposia, 20* : 427-55.

Baggerman, B. (1968) 'Hormonal control of reproductive and parental behaviour in fishes' in E.J.W. Barrington & C.B. Jørgensen (eds.) *Perspectives in Endocrinology,* Academic Press, New York, pp. 351-404.

Baggerman, B. (1980) 'Photoperiodic and endogenous control of the annual reproductive cycle in teleost fishes.' *NATO Advanced Study Institute, Series (A), 35* : 535-67.

Bagnara, J.T. and Hadley, M.E. (1973) *Chromatophores and colour change.* Prentice Hall Inc., Englewood Cliffs, N.J.

Bail, P.Y. le and Breton, B. (1981) 'Rapid determination of the sex of puberal salmonid fish by a technique of immunoagglutination.' *Aquaculture, 22* : 367-75.

Bailey, J.R. and Randall, D.J. (1981) 'Renal perfusion pressure and renin secretion in the rainbow trout, *Salmo gairdneri.*' *Canadian Journal of Zoology, 59* : 1220-6.

Ballintijn, C.M. (1972) 'Efficiency, mechanics and motor control of fish respiration.' *Respiration Physiology, 14* : 125-41.

Ballintijn, C.M. (1982) 'Neural control of respiration in fishes and mammals' in A.D.F. Addink & N. Spronk (eds.) *Exogenous & Endogenous Influences on Metabolic and Neural Control,* Pergamon Press, Oxford, pp. 127-40.

Ballintijn, C.M. and Roberts, J.L. (1976) 'Neural control and proprioceptive load matching in reflex respiratory movements of fishes.' *Federation Proceedings, 35* : 1983-91.

Ballintijn, C.M., Luiten, P.G.M. and Jüch, P.J.W. (1979) 'Respiratory neuron activity in the mesencephalon, diencephalon and cerebellum of the carp.' *Journal of Comparative Physiology, A, 133* : 131-9.

Balon, E.K. (1975). 'Reproductive guilds of fishes : a proposal and definition.' *Journal of the Fisheries Research Board of Canada, 32* : 821-64.

Bamford, O.S. (1974). 'Oxygen reception in the rainbow trout *(Salmo gairdneri).*' *Comparative Biochemistry & Physiology, 48A* : 69-76.

Banks & Guthrie 1982. Unpublished observations

Barlow, G.W. and Ballin, P.J. (1976) 'Predicting and assessing dominance from size and colouration in the polychromatic Midas cichlid.' *Animal Behaviour, 24* : 793-813.

Baumeister, H.G. (1973) 'Lampbrush chromosomes and RNA synthesis during early oogenesis of *Brachydanio rerio* (Cyprinidae, Teleostei).' *Zeitschrift für Zellforschung und Mikroskopische Anatomie, 145* : 145-50.

Baumgarten, H.G. (1977) 'Verkommen und Verteilung adrenerger Nervenfasern in Darm der Schleie *(Tinca vulgaris* Cuv.)' *Zeitschrift für Zellforschung und Mikroskopische Anatomie, 76* : 248-59.

Baumgarten, H.G., Bjorklünd, A. Nobin, A. and Rosengren, E. (1973) 'Evidence for the existence of serotonin-, dopamine-, and noradrenaline-containing fibres in the gut of *Lampetra fluviatilis.*' *Zeitschrift für Zellforschung und Mikroskopische Anatomie, 141* : 33-54.

Bellamy, D. (1968) 'Metabolism of the red piranha *(Rooseveltiella nattererii)* in relation to feeding behaviour.' *Comparative Biochemistry and Physiology, 25* : 343-7.

Bellamy, D. and Chester Jones, I. (1961) 'Studies on *Myxine glutinosa* — 1. The chemical composition of the tissues.' *Comparative Biochemistry and Physiology, 3* : 175-83.

Bennett, M.V.L. (1974) 'Flexibility and rigidity in electrotonically coupled systems' in *Synaptic transmission and neuronal interaction*, Raven Press, New York, pp. 153-78.

Bergman, H.L., Olson, K.R. and Fromm, P.O. (1974) 'The effects of vasoactive agents on the functional surface area of the isolated perfused gills of rainbow trout.' *Journal of Comparative Physiology, 94* : 267-86.

Beyer, C. (1979) *Endocrine Control of Sexual Behavior.* Raven Press, New York.

Beyenbach, K.W. (1982) 'Direct demonstration of fluid secretion by glomerular renal tubules in a marine teleost.' *Nature 299* : 54-6.

Bhatt, S.D. and Khanna, S.S. (1976) 'Histopathological effects of glucagon in a freshwater teleost, *Clarias batrachus* L. Part 1. The endocrine pancreas, blood glucose and tissue glycogen.' *Indian Journal of Experimental Biology, 14* : 145-9.

Billard, R. (1978) 'Testicular feedback on the hypothalamo-pituitary axis in rainbow trout *(Salmo gairdneri R.).*' *Annales de Biologie Animale Biochimie Biophysique, 18* : 813-18.

Billard, R. and Breton, B. (1977) 'Sensibilité à la température des différentes étapes de la reproduction chez la truite arc-en-ciel.' *Cahiers du laboratoire de Montereau, 5* : 5-24.

Billard, R., Breton, B. and Jalabert, B. (1971) 'La production spermatogénétique chez la truite.' *Annales de Biologie Animale Biochimie Biophysique, 11* : 190-212.

Billard, R., Bry, C. and Gillet, C. (1981) 'Stress, environment and reproduction in teleost fish' in A.D. Pickering (ed.), *Stress in Fish*, Academic Press, New York, pp. 185-208.

Billard, R., Breton, B., Fostier, A., Jalabert, B. and Weil, C. (1978) 'Endocrine control of the teleost reproductive cycle and its relation to external factors : salmonid and cyprinid models' in P.J. Galliard & H.H. Boer (eds.), *Comparative Endocrinology*, Elsevier North, Amsterdam pp 37-48.

Bilton, H.T. and Robins, G.L. (1973) 'The effects of starvation and subsequent feeding on survival and growth of Fulton Channel sockeye salmon *(Oncorhynchus nerka).*' *Journal of the Fisheries Research Board of Canada, 30* : 1-5.

Birnbaum, M.J., Schulz, J. and Fain, J.N. (1976) 'Hormone stimulated glycogenolysis in isolated goldfish hepatocytes.' *American Journal of Physiology, 231* : 191-7.

Blinkov, S.M. and Glezer, I.I. (1968) *The Human Brain in Figures and Tables: a Quantitative Handbook*, Basis Books Inc, New York.

Bloom, G., Östlund, E., Euler, U.S. von., Lishajko, F., Ritzén, M. and Adams-Ray, J. (1961) 'Studies on catecholamine containing granules of specific cells in cyclostome hearts.' *Acta Physiologica Scandinavica, 53* : *Supplement 185*, 1-35.

Blum, V. (1974) 'Die Rolle des Prolaktins bei der Cichlidenbrutpflege.' *Fortschrifte Zoologie, 22* : 310-33.

Bohemen, Ch. G. van and Lambert, J.G.D. (1981) 'Estrogen synthesis in relation to estrone, estradiol and vitellogenin plasma levels during the reproductive cycle of the female rainbow trout, *Salmo gairdneri.*' *General and Comparative Endocrinology, 45* : 105-14.

Bohemen, Ch. G. van, Lambert, J.G.D. and Peute, J. (1981) 'Annual changes in plasma and liver in relation to vitellogenesis in the female rainbow trout, *Salmo gairdneri.*' *General and Comparative Endocrinology, 44* : 94-107.

Bohr, C. (1894) 'The influence of section of the vagus nerve on disengagement of gases in the air-bladder of fishes.' *Journal of Physiology, 15* : 494-500.

Booth, D.A. (1978) 'Prediction of feeding behaviour from energy flows in the rat' in D.A. Booth (ed.), *Hunger models : computable theory of feeding control,* Academic Press, New York, pp. 227-78.

Booth, J.H. (1979) 'The effects of oxygen supply, epinephrine and acetylcholine on the distribution of blood flow in trout gills.' *Journal of Experimental Biology, 83* : 31-9.

Borsook, D., Woolf, C.J., Vellet, D., Abramson, I.S. and Shapiro, C.M. (1981) 'Pituitary peptides and memory in fish' in P.R. Laming (ed.) *Brain Mechanisms of behaviour in lower vertebrates,* University Press, Cambridge, pp. 185-92.

Bouffard, R.E. (1979) 'The Role of Prostaglandins During Sexual Maturation, Ovulation and Spermiation in the Goldfish, *Carassius auratus.*' Unpublished M.Sc. Thesis, University of British Columbia.

Boylan, J.W. (1972) 'Model for passive urea reabsorption in the elasmobranch kidney.' *Comparative Biochemistry and Physiology, 42A* : 27-30.

Breder, C.M. and Rosen, D.E. (1966) *Modes of Reproduction in Fishes,* Natural History Press, New York.

Breton, B. and Billard, R. (1977) 'Effects of photoperiod and temperature on plasma gonadotrophin and spermatogenesis in the rainbow trout, *Salmo gairdneri R.*' *Annales de Biologie Animale Biochimie Biophysique, 17* : 331-40.

Brinn, J.E., Jr. (1973) 'The Pancreatic Islets of Bony Fishes.' *American Zoologist, 13* : 653-65.

Bromage, N.R., Whitehead, C. and Breton, B. (1982) 'Relationships between serum levels of gonadotrophin, oestradiol-17β and vitellogenin in the control of ovarian development in the rainbow trout: II The effects of alterations in environmental photoperiod.' *General and Comparative Endocrinology, 47* : 366-76.

Brown, L.A. and Richards, R.M. (1979) 'Surgical gonadectomy of fish: a technique for veterinary surgeons.' *Veterinary Record, 104* : 215.

Brown, J.A., Jackson, B.A., Oliver, J.A. and Henderson, I.W. (1978) 'Single nephron filtration rate (SNGFR) in the trout, *Salmo gairdneri.* Validation of the use of ferrocyanide and the effect of environmental salinity.' *Pflügers Archiv. 377* : 101-8.

Brown, J.A., Oliver, J.A., Henderson, I.W. and Jackson, B.A. (1980) 'Angiotensin and single nephron glomerular filtration in the trout, *Salmo gairdneri.*' *American Journal of Physiology, 239* : R509-14.

Brown, J.A. and Oliver, J.A. (1983) 'The renin-angiotensin system and single nephron glomerular structure and function in the trout, *Salmo gairdneri.*' *Symposium Proceedings of the Ninth International Symposium on Comparative Endocrinology, Hong Kong, 1981.* (In press).

Bry, C. (1981) 'Temporal aspects of macroscopic changes in rainbow trout *(Salmo gairdneri)* oocytes before ovulation, and of ova fertility during the post-ovulation period: effect of treatment with 17α-hydroxy-20β dihydroprogesterone.' *Aquaculture, 24* : 153-60.

Budker, P. (1958) 'La viviparité chez les sélaciens' in P. Grassé (ed.) *Traité de Zoologie, 13,* Fasc. 2, 1755-90, Masson, Paris.

Budker, P. (1971) *The life of sharks,* Columbia University Press, New York.

Bullock, T.H., Hamstra, R.H. and Scheich, H. (1972) 'The jamming avoidance of high frequency electric fish, I.' *Journal of Comparative Physiology, 77* : 1-22.

Burkhardt, D.A. (1966) 'The goldfish electroretinogram : relation between photopic spectral sensitivity functions and cone absorption spectra.' *Vision Research, 6* : 517-32.

Burkhardt, D.A. (1977) 'Responses and receptive field organisation of cones in perch retinas.' *Journal of Neurophysiology, 40* : 53-62.

Burnstock, G. (1969) 'Evolution of the autonomic innervation of visceral and cardiovascular systems in vertebrates.' *Pharmacological Reviews, 21* : 247-324.

Burnstock, G. (1981) 'Review lecture : Neurotransmitters and trophic factors in the autonomic nervous system.' *Journal of Physiology, 313* : 1-35.

Busack, C.A. and Gall, G.A.E. (1980) 'Ancestry of artificially propagated California rainbow trout strains.' *California Fish and Game, 66* : 17-24.

Buser, P. (1955) 'Analyses des responses electriques du lobe optique à la stimulation de la voie visuelle chez quelque vertebrés inferieures.' Unpublished Ph.D. Thesis, University of Paris, France.

Buss, K. (1980) 'Photoperiod control for brood trout. Manipulating the spawning season to meet production requirements.' *Aquaculture Magazine, 6* : 45-8.

Butler, P.J. (1976) 'Gas exchange' in J. Bligh, J.L. Cloudsley-Thompson & A.G. Macdonald (eds.) *Environmental Physiology of Animals*, Blackwell, Oxford, pp. 164-95.

Butler, P.J. and Taylor, E.W. (1975) 'The effect of progressive hypoxia on respiration in the dogfish *(Scyliorhinus canicula)* at different seasonal temperatures.' *The Journal of Experimental Biology, 63* : 117-30.

Butler, P.J., Taylor, E.W. and Short, S. (1977) 'The effect of sectioning cranial nerves V, VII, IX and X on the cardiac response of the dogfish *Scyliorhinus canicula* to environmental hypoxia.' *The Journal of Experimental Biology, 69* : 233-45.

Butler, P.J., Taylor, E.W., Capra, M.F., and Davison, W. (1978) 'The effect of hypoxia on the levels of circulating catecholamines in the dogfish, *Scyliorhinus canicula.' Journal of Comparative Physiology B : 127*, 325-30.

Butler, P.J., Taylor, E.W. and Davison, W. (1979) 'The effect of long term moderate hypoxia on acid-base balance, plasma catecholamines and possible anaerobic end products in the unrestrained dogfish, *Scyliorhinus canicula.' Journal of Comparative Physiology B , 132* : 297-303.

Bye, V.J. and Jones, A. (1981) 'Sex control — a method for improving productivity in turbot farming.' *Fish Farming International, 8* : 31-2.

Bye, V.J. and Lincoln, R.F. (1981) 'Get rid of the males and let the females prosper.' *Fish Farmer, 4* : 22-4.

Callard, G.V., Petro, Z. and Ryan, K.J. (1978a) 'Conversion of androgen to oestrogen and other steroids in the vertebrate brain.' *American Zoologist, 18* : 511-24.

Callard, G.V., Petro, Z. and Ryan, K.J. (1978b) 'Phylogenetic distribution of aromatase and other androgen converting enzymes in the central nervous system.' *Endocrinology, 103* : 2283-90.

Callard, G.V., Petro, Z. and Ryan, K.J. (1981) 'Oestrogen synthesis *in vitro* and *in vivo* in the brain of a marine teleost *(Myoxocephalus).' General Comparative Endocrinology, 43* : 243-55.

Cameron, J.N. (1980) 'Body fluid pools, kidney function and acid-base regulation in the freshwater catfish, *Ictalurus punctatus.' Journal of Experimental Biology, 86* : 171-85.

Campbell, C.M. and Idler, D.R. (1980) 'Characterisation of an estradiol-induced protein from rainbow trout serum as vitellogenin by the composition and radioimmunological cross-reactivity to ovarian yolk fractions.' *Biology of Reproduction, 22* : 605-17.

Campbell, C.M., Fostier, B., Jalabert, B. and Truscott, B. (1980) 'Identification and quantification of steroids in the serum of spermiating or ovulating rainbow trout.' *Journal of Endocrinology, 85* : 371-8.

Campbell, G. (1970a) 'Autonomic nervous supply to effector tissues' in E.

Bülbring, A.F. Brading, A.W. Jones & T. Tomita (eds.) *Smooth Muscle*, London, E. Arnold, pp. 451-95.

Campbell, G. (1970b) 'Autonomic nervous systems' in W.S. Hoar & D.J. Randall (eds.) *Fish Physiology, Vol. IV*, Academic Press, New York, pp. 109-32.

Campbell, G. and Burnstock, G. (1968) 'Comparative Physiology of gastrointestinal motility' in C.F. Code (ed.) *Handbook of Physiology, Sec. 6, Part IV (Alimentary Canal; Motility)*, American Physiology Society, Washington, U.S.A. pp. 2213-66.

Campbell, G. and Gannon, B.J. (1976) 'The splanchnic nerve supply to the stomach of the trout, *Salmo trutta and S. gairdneri.' Comparative Biochemistry & Physiology, 55C*, 51-3.

Campbell, G. and Gibbins, I.L. (1979) 'Nonadrenergic, noncholinergic transmission in the autonomic nervous system: Purinergic nerves' in S. Kalsner (ed.) *Trends in Autonomic Pharmacology, Vol. 1*, Urban & Schwarzenburg, Munich, pp. 103-43.

Capra, M.F. (1976) 'Cardio-respiratory relationships during "eupnoea" and respiratory arrhythmias in the Port Jackson shark, *Heterodontus portusjacksoni.' Comparative Biochemistry and Physiology, 53A* : 259-62.

Capra, M.F. and Satchell, G.H. (1977) 'Adrenergic and cholinergic responses of the isolated saline-perfused heart of the elasmobranch fish, *Squalus acanthias.' General Pharmacology, 8* : 56-65.

Carlin, B. (1968) 'Salmon conservation, tagging experiments and migrations of salmon in Sweden.' *Lecture Series, the Atlantic Salmon Association*, Montreal, Canada.

Castilla, C. and Murat, J.C. (1957) 'Effets de l'insuline sur le métabolisme protéique dans le foie de carpe.' *Comptes Rendus des Séances, Societé de Biologie et de ses Filiales, 169* : 1605-9.

Chan, D.K.O. and Woo, N.Y.S. (1978) 'Effect of glucagon on the metabolism of the Japanese eel.' *General & Comparative Endocrinology, 35* : 216-25.

Chevassus, B. (1979) 'Hybridisation in salmonids : results and perspectives.' *Aquaculture, 17* : 113-28.

Chieffi, G. (1961) 'La luteogenesi nei Selaci ovovivipari. Richerche istologiche e istochimiche' in *Torpedo marmorata* e *Torpedo ocellata.' Pubb. staz. Zool. Napoli, 32* : 145-66.

Chieffi, G. (1962) 'Endocrine aspects of reproduction in elasmobranch fishes.' *General & Comparative Endocrinology, Supplement 1* : 275-85.

Chieffi, G. (1967) 'The reproductive system of elasmobranchs : developmental and endocrinological aspects' in P.W. Gilbert, R.F. Mathewson and D.P. Hall (eds.) *Sharks, Skates and Rays*. Johns Hopkins Press, Baltimore, pp. 553-80.

Chieffi, G. and Lupo di Prisco, C. (1961) 'Identification of estradiol-17β testosterone and its precursors from *Scyliorhinus stellaris* testes.' *Nature, London, 1905* : 169-70.

Chieffi, G. and Lupo di Prisco, C. (1963) 'Identification of sex hormones in the ovarian extract of *Torpedo marmorata* and *Bufo vulgaris.' General and Comparative Endocrinology, 3* : 149-52.

Chieffi, G., Della Corte, F. and Botte, V. (1961) 'Osservazioni sul tessuto interstiziale del testicolo dei Selacei.' *Bolletino di zoologia, 28* : 211-17.

Chourrot, D. (1980) 'Thermal induction of diploid gynogenesis and triploidy in the eggs of the rainbow trout *(Salmo gairdneri R.).' Reproduction Nutrition Développement, 20* : 727-33.

Churchill, P.C., Malvin, R.L., Churchill, M.C. and McDonald, F.D. (1979) 'Renal function in *Lophius americanus*. Effects of angiotensin II.' *American Journal of Physiology, 236* : R297-301.

Clark, R.S. (1922) 'Rays and skates (Raiae) No.1. Egg capsules and young.'

Journal of the Marine Biological Association, U.K. 12 : 577-643.

Colgan, P. (1973) 'Motivational analysis of fish feeding.' *Behaviour, 45* : 38-67.

Collenot, G. and Ozon, R. (1964) 'Mises en évidence biochimique et histochimique d'une Δ⁵,3β hydroxystéroid déshydrogénase dans le testicule de *Scyliorhinus canicula L.' Bulletin de la Société zoologique de France, 89* : 577-87.

Compagno, L.J.V. (1977) 'Phyletic relationships of living sharks and rays.' *American Zoologist, 17* : 303-22.

Cook, A.F., Stacey, N.E. and Peter, R.E. (1980) 'Periovulatory changes in serum cortisol levels in the goldfish, *Carassius auratus.' General and Comparative Endocrinology, 40* : 507-10.

Cooper, J.C. and Scholz, A.T. (1976) 'Homing of artificially imprinted steelhead trout.' *Journal of Fisheries Research Board of Canada, 33* : 826-9.

Cooper, J.C., Scholz, A.T., Horrall, R.M., Hasler, A.D. and Madison, D.M. (1976) 'Experimental confirmation of the olfactory hypothesis with artificially imprinted homing coho salmon, *(Oncorhynchus kisutch).' Journal of Fisheries Research Board of Canada, 33* : 703-10.

Cooperstein, S.J. and Lazarow, A. (1969) 'Uptake of glucose by islet of Langerhans and other tissues of the toadfish, *Opsanus tau.' American Journal of Physiology, 217* : 1784-88.

Corson, B.W. (1955) 'Four years progress in the use of artificially controlled light to induce early spawning of brook trout.' *Progressive Fish Culturist, 17* : 99-102.

Coupland, R.E. (1965) *The natural history of the chromaffin cell*, Longmans, London.

Cowey, C.B. and Sargent, J.R. (1979) 'Nutrition' in W.S. Hoar, D.J. Randall & J.R. Brett (eds.) *Fish Physiology, Volume VIII*, Academic Press, New York, pp. 1-69.

Cowey, C.B. De La Higuera, M. and Adron, J.W. (1977) 'The effect of dietary composition, and of insulin on gluconeogenesis in rainbow trout *(Salmo gairdneri).' British Journal of Nutrition, 38* : 385-95.

Cowey, C.B., Knox, D., Walton, M.J. and Adron, J.W. (1977) 'The regulation of gluconeogenesis by diet and insulin in rainbow trout, *(Salmo gairdneri).' British Journal of Nutrition, 38* : 463-70.

Craik, J.C.A. (1978a) 'Plasma levels of vitellogenin in the elasmobranch *Scyliorhinus canicula L.* (lesser spotted dogfish).' *Comparative Biochemistry & Physiology, 60B* : 9-18.

Craik, J.C.A. (1978b) 'Kinetic studies of vitellogenin metabolism in the elasmobranch *Schyliorhinus canicula L.' Comparative Biochemistry & Physiology, 61A* : 355-61.

Craik, J.C.A. (1978c) 'An annual cycle of vitellogenesis in the elasmobranch, *Scyliorhinus canicula.' Journal of the Marine Biological Association, U.K. 58* : 719-26.

Craik, J.C.A. (1978d) 'The effects of oestrogen treatment on certain plasma constituents associated with vitellogenesis in the elasmobranch, *Scyliorhinus canicula L.' General and Comparative Endocrinology, 35* : 455-64.

Crews, D. and Silver, R. (1980) 'Reproductive physiology and behavior interactions in nonmammalian vertebrates' in R.W. Goy & D.W. Pfaff (eds.) *Handbook of Behavioural Neurobiology*, Plenum Press, New York.

Crim, L.W., Peter, R.E. and Billard, R. (1981) 'Onset of gonadotrophin accumulation in the immature trout pituitary gland in response to estrogen or aromatizable androgen steroid hormones.' *General and Comparative Endocrinology, 44* :374-81.

Cronly-Dillon, J. and Sharma, S.C. (1968) 'Effect of season and sex on the photopic spectral sensitivity in the three-spined stickleback.' *Journal of Experimental Biology, 49* : 679-87.

Crow, R.T. and Liley, N.R. (1979) 'A sexual pheromone in the guppy, *Poecilia reticulata* (Peters).' *Canadian Journal of Zoology, 57* : 184-8.

Daughaday, W.H., Herington, A.C. and Phillips, L.S. (1975) 'The regulation of growth by endocrines.' *Annual Review of Physiology, 37* : 211-44.

Davies, D.T. and Rankin, J.C. (1973) 'Adrenergic receptors and vascular responses to catecholamines of perfused dogfish gills.' *Comparative and General Pharmacology, 4* : 139-47.

Davis, J.O. and Freeman, R.H. (1976) 'Mechanisms regulating renin release.' *Physiological Reviews, 56* : 3-56.

Davis, R.E. (1975) 'Readiness to display in the paradise fish *Macropodus opercularis* L., Belontiidae: the problem of general and specific effects of social isolation.' *Behavioural Biology, 15* : 419-33.

Davis, R.E. and Kassel, J. (1975) 'The ontogeny of agonistic behaviour and the onset of sexual maturation in the paradise fish, *Macropodus opercularis (L.).*' *Behavioural Biology, 14* : 31-9.

Davis, R.E., Mitchell, M. and Dolson, L. (1975) 'The effects of methallibure on conspecific visual reinforcement, social display, frequency and spawning in the paradise fish, *Macropodus opercularis (L.).*' *Physiology of Behaviour, 17* : 47-52.

Daw, N.W. (1968) 'Colour coded ganglion cells in the goldfish retina.' *Journal of Physiology, 197* : 567-92.

Daxboeck, C. and Holeton, G.F. (1978) 'Oxygen receptors in the rainbow trout, *Salmo gairdneri.*' *Canadian Journal of Zoology, 56* : 1254-9.

deBruin, J.P.C. (1980) 'Telencephalon and behaviour in teleost fish' in S.O.E. Ebbesson (ed.), *Comparative Neurology of the Telencephalon*, Plenum, New York.

deBruin, J.P.C. (1982) 'Neural correlates of motivated behaviour in fish' in P. Ewart, R. Capranica & I. Ingle (eds.) *Advances in Vertebrate Neuroethology*, Plenum, London. (In press).

deVlaming, V.L. (1974) 'Environmental and endocrine control of teleost reproduction' in C.B. Schreck (ed.), *Control of Sex in Fishes*, Virginia Polytechnic Institute, Blacksburg, pp. 13-83.

deVlaming, V.L. and Pardo, R.J. (1975) '*In vitro* effects of insulin on liver lipid and carbohydrate metabolism in the teleost *Notemigonus crysoleucas.*' *Comparative Biochemistry & Physiology, 51B* : 489-97.

deVlaming, V.L., Wiley, H.S., Delahunty, G. and Wallace, R.A. (1980) 'Goldfish *(Carassius auratus)* vitellogenin : Induction, isolation, properties and relationship to yolk proteins.' *Comparative Biochemistry and Physiology, 67B* : 613 23.

Deery, D.J. (1974) 'Determination by radioimmunoassay of the luteinising hormone-releasing hormone (LHRH) content of the hypothalamus of the rat and some lower vertebrates.' *General and Comparative Endocrinology, 24* : 280-5.

Deery, D.J. and Jones, A.C. (1974) 'Effects of hypothalamic extracts, neurotransmitters and synthetic hypothalamic releasing hormones on adenylyl cyclase activity in the lobes of the pituitary of the dogfish *(Scyliorhinus canicula L.).*' *Journal of Endocrinology, 64* : 49-57.

Dejours, P. (1981) *Principles of Comparative Respiratory Physiology*, (2nd edn), Elsevier, Amsterdam.

Delaney, R.G., Lahiri, S., Hamilton, R. and Fishman, A.P. (1977) 'Acid-base balance and plasma composition in the estivating lungfish *(Protopterus).*' *American Journal of Physiology, 232* : R10-17.

Della Corte, F., Botte, V. and Chieffi, G. (1961) 'Ricerca istochimica della steroide-3β-olodeidrogenasi nel testicolo di *Torpedo marmorata* Risso e di *Scyliorhinus stellaris (L.).*' *Atti Soc. Pelorit. Sci. Fis. Mat. Natur.* 7 : 393-7.

Demski, L.S. (1973) 'Feeding and aggressive behaviour evoked by hypothalamic stimulation in a cichlid fish.' *Comparative Biochemistry and Physiology, 44A* : 685-92.

Demski, L.S. (1981) 'Hypothalamic mechanisms of feeding in fishes' in P.R. Laming (ed.), *Brain Mechanisms of Behaviour in Fishes, Society for Experimental Biology, Seminar Series 9,* University Press, Cambridge, pp. 225-238.

Demski, L.S. and Knigge, K.M. (1971) 'The telencephalon and the hypothalamus of the blue gill, *(Lepomis macrochirus)*: evoked feeding, aggressive and reproductive behaviours' *Journal of Comparative Neurology, 143* : 1-16.

Demski, L.S. and Hornby, P.J. (1982) 'Hormonal control of fish reproductive behaviour : brain-gonadal steroid interactions.' *Canadian Journal of Fisheries and Aquatic Sciences, 39 (1)* : 36-47.

Diakow, C. and Nemiroff, A. (1981) 'Vasotocin, prostaglandin, and female reproductive behavior in the frog, *Rana pipiens.*' *Hormones and Behavior, 15* : 86-93.

Diakow, D. and Raimondi, D. (1981) 'Physiology of *Rana pipiens* reproductive behavior : a proposed mechanism for inhibition of the release call.' *American Zoologist, 21* : 295-304.

Dobson, S. (1974) 'Endocrine control of reproduction in the male *Scyliorhinus canicula.*' Unpublished Ph.D. thesis. University of Wales, U.K.

Dobson, S. and Dodd, J.M. (1977a) 'Endocrine control of the testis in the dogfish *Scyliorhinus canicula* I. Effects of partial hypophysectomy on gravimetric, hormonal and biochemical aspects of testis function.' *General and Comparative Endocrinology, 32* : 41-52.

Dobson, S. and Dodd, J.M. (1977b) 'Endocrine control of the testis in the dogfish *Scyliorhinus canicula L.* II. Histological and ultrastructural changes in the testis after partial hypophysectomy (ventral lobectomy).' *General and Comparative Endocrinology, 32* : 53-71.

Dodd, J.M. (1960) 'Genetic and environmental aspects of sex determination in cold-blooded vertebrates.' *Mem. Soc. Endocrinol., 7* : 17-44.

Dodd, J.M. (1972) 'Ovarian control of cyclostomes and elasmobranchs.' *American Zoologist, 12* : 325-39.

Dodd, J.M. (1975) 'The hormones of sex and reproduction and their effects in fish and lower chordates : 20 years on.' *American Zoologist, 15* : Supplement 137-71.

Dodd, J.M. (1977) 'The structure of the ovary of non-mammalian vertebrates' in S. Zuckerman and B. Weir (eds.), *The Ovary,* 2nd edn., Vol. 1, pp. 219-63.

Dodd, J.M. (1983) 'Reproduction in cartilaginous fishes' in W.S. Hoar and D.J. Randall (eds.) *Fish Physiology,* Academic Press, New York (In press).

Dodd, J.M., Evennett, P.J. and Goddard, C.K. (1960) 'Reproductive endocrinology in cyclostomes and elasmobranchs.' *Symposium of the Zoological Society, London, 1* : 77-103.

Dodd, J.M. and Goddard, C.K. (1961) 'Some effects of oestradiol benzoate on the reproductive ducts of *Scyliorhinus canicula.*' *Proceedings of the Zoological Society, London, 137* : 325-32.

Dodd, J.M. and Sumpter, J.P. (1983) 'Reproductive cycles of cyclostomes, elasmobranchs and bony fishes' in G.E. Lamming (ed.), *Marshall's Physiology of Reproduction,* 4th edn., Chapter 1, Churchill Livingstone, Edinburgh, (In press).

Dodd, M.H.I. and Dodd, J.M. (1980) 'Ultrastructure of the ovarian follicle of the dogfish, *Scyliorhinus canicula L.' General and Comparative Endocrinology, 40* : 330-1.

Dodd, M.H.I. and Duggan, R.T. (1982) 'Trawl stress ovulation in the dogfish, *Scyliorhinus canicula.' General and Comparative Endocrinology, 46* : 392.

Donaldson, E.M., Hunter, G.A. and Dye, H.M. (1981) 'Induced ovulation in coho salmon *(O. kisutch)*. III Preliminary study on the use of the antiestrogen Tamoxifen.' *Aquaculture, 26* : 143-54.

Donaldson, E.M. and Hunter, G.A. (1982) 'Sex control in fishes with particular reference to salmonids.' *Canadian Journal of Fisheries and Aquatic Sciences, 39* : 99-110.

Donaldson, R. and Allen, G.H. (1957) 'Return of silver salmon, *Oncorhyncus kisutch* (Walbaum), to point of release.' *Transactions of the American Fisheries Society, 87* : 13-22.

Døving, K.B. (1979) 'The secret cues for salmon navigation' in *Research in Norway, 1979*, pp. 45-50.

Døving, K.B. (1982) 'Behaviour patterns in the cod released by electrical stimulation of olfactory bundlets' in P. Ewert, R. Capranica and I. Ingle (eds.) *Advances in Vertebrate Neuroethology*, Plenum, London.

Døving, K.B., Nordeng, H. and Oakley, B. (1974) 'Single unit discrimination of fish odors released by char *(Salmo alpinus L.)* populations.' *Comparative Biochemistry & Physiology, 47A* : 1051-63.

Døving, K.B., Selset, R. and Thommesen, G. (1980) 'Olfactory sensitivity to bile acids in salmonid fishes.' *Acta Physiologica Scandinavica, 1980* : 24-31.

Dubois, M.P., Billard, R., Breton, B. and Peter, R.E. (1979) 'Comparative distribution of somatostatin, LH—RH, neurophysin and alphaendorphin in the rainbow trout : an immunocytological study.' *General & Comparative Endocrinology, 37* : 220-32.

Ebbesson, S. (1980) 'The Parcellation theory and its relation to interspecific variability in brain organization, evolutionary and ontogenetic development, and neuronal plasticity.' *Cell Tissue Research, 213* : 179-212.

Egami, N. (1959) 'Preliminary note on the induction of the spawning reflex and oviposition in *Oryzias latipes* by the administration of neurohypophyseal substances.' *Annotationes Zoologicae Japonensis, 32* : 13-17.

Elger, M. and Hentschel, H. (1981) 'Präglomeruläre sphinktere im opistonephros der regnbogenforelle, *Salmo gairdneri.' Verhandlungen der Deutschen Zoologischen Gesselschaft (1981)* : 218.

Elliott, J.M. (1975) 'Number of meals in a day, maximum weight of food consumed in a day and maximum rate of feeding in brown trout, *Salmo trutta L.' Freshwater Biology, 5* : 287-303.

Elliott, J.M. and Persson, L. (1978) 'The estimation of daily rates of food consumption for fish.' *Journal of Animal Ecology, 47* : 977-93.

Elul, R. (1972) 'The genesis of the E.E.G.' *International Review of Neurobiology, 15* : 228-72.

Epple, A. (1969) 'The endocrine pancreas' in W.S. Hoar and D. J. Randall (eds.) *Fish Physiology, Vol. 2*, Academic Press, New York, pp. 275-310.

Epple, A. and Lewis, T.L. (1975) 'The effect of pancreatectomy on the survival of *Anguilla rostrata* in different salinities.' *Journal of Experimental Zoology, 193* : 457-62.

Epple, A. and Lewis, T.L. (1977) 'Metabolic effects of pancreatectomy and hypophysectomy in the yellow American eel, *Anguilla rostrata* LeSeuer.' *General and Comparative Endocrinology, 32* : 294-315.

Epple, A. and Kocsis, J.J. (1980) 'Effects of pancreatectomy on tissue taurine of the American eel under varying osmotic and nutritional conditions.' *Comparative Biochemistry and Physiology, 65A* : 139-42.

Erdahl, D. and Graham, E.F. (1980) 'Preservation of gametes of freshwater fish.' *Proceedings of the 9th International Congress on Animal Reproduction and Artificial Insemination, 2. Round Tables. Madrid, 16-20 June 1980.* pp. 317-26.

Eriksson, L.O. and Lundquist, H. (1980) 'Photoperiod entrains ripening by its differential effect in salmon.' *Naturwissenschaften 67* : 202-3.

Escaffre, A.M., Petit, J. and Billard, R. (1977) 'Évolution de la quantité d'ovules récoltés et conservation de leur aptitude à être fécondés au cours de la période post-ovulatoire chez la truite arc-en-ciel.' *Bulletin francaise de Pisciculture 265* : 134-42.

Euler, U.S. von. and Fänge, R. (1961) 'Catecholamines in nerves and organs of *Myxine glutinosa, Squalus acanthias & Gadus callarias.*' *General and Comparative Endocrinology, 1* : 191-4.

Evans, D.H. (1982) 'Salt and water exchange across vertebrate gills' in D.F. Houlihan, J.C. Rankin and T. J. Shuttleworth (eds.), *Gills,* University Press, Cambridge, pp. 149-71.

Ezeasor, D.N. (1979) 'Ultrastructural observations on the submucous plexus of the large intestine of the rainbow trout *(Salmo gairdneri,* Rich.).' *Z. mikorsk-anat Forsch, Lepizig, 93* : 803-12.

Falck, B., Münzing, J. and Rosengren, A.-M. (1969) 'Adrenergic nerves to the dermal melanophores of the rainbow trout *Salmo gairdneri.*' *Zeitschrift fur Zellforschung und Microscopische Anatomie, 99* : 430-4.

Falck, B., Von Mecklenburg, C., Myhrberg, H. and Persson, H. (1966) 'Studies on adrenergic and cholinergic receptors in the isolated hearts of *Lampetra fluviatalis* (Cyclostomata) and *Pleuronectes* (Teleostei).' *Acta Physiologica Scandinavica 68* : 64-71.

Fänge, R. (1953) 'The mechanisms of gas transport in the euphysoclist swim-bladder.' *Acta Physiologica Scandinavica 30* : *Supplement 110* : 1-133.

Fänge, R. (1963) 'Structure and function of the excretory organs of myxinoids' in A. Brodal and R. Fänge (eds.), *The Biology of Myxine,* Universitetsforlaget, Oslo, pp. 516-29.

Fänge, R. (1976) 'Gas exchange in the swimbladder' in G.M. Hughes (ed.), *Respiration in amphibious vertebrates,* Academic Press, New York, pp. 189-211.

Fänge, R. and Östlund, E. (1954) 'The effects of adrenaline, noradrenaline, tyramine and other drugs on the isolated heart from marine vertebrates and a cephalopod *(Eledone cirrosa).*' *Acta Zoologica (Stockholm), 35* : 289-305.

Fänge, R. and Johnels, A.G. (1958) 'An autonomic nerve plexus control of the gall bladder in *Myxine.*' *Acta Zoologica (Stockholm), 39* : 1-8.

Fänge, R., Johnels, A.G. and Enger, P.S. (1963) 'The autonomic nervous system' in A. Brodal and R. Fänge (eds.), *The biology of Myxine,* Universitetsforlaget, Oslo, pp. 122-36.

Fänge, R. and Grove, D.J. (1979) 'Digestion' in W.S. Hoar, D.J. Randall and J. Brett (eds.), *Fish Physiology, Vol. VIII,* Academic Press, New York, pp. 161-260.

Fänge, R. and Mattisson, A. (1981) 'The lymphomyeloid (haemopoietic) system of the atlantic nurse shark, *Ginglymostoma cirratum.*' *Biol. Bull. Mar. Biol. Lab., Woods Hole, 160* : 240-9.

Fänge, R. and Holmgren, S. (1982) 'Choline acetyltransferase activity in the fish swimbladder.' *Journal of Comparative Physiology, 146B* : 57-61.

Fenwick, J.C. (1970) 'The pineal organ' in W.S. Hoar and D.J. Randall (eds.), *Fish Physiology, Vol. IV,* Academic Press, New York.

Fernald, R. (1976) 'The effect of testosterone on the behaviour and colouration of adult male cichlid fish *(Haplochromis burtoni* Gunther).' *Hormone Research,* 7: 172-8.

Fiedler, K. (1974) 'Hormonalle Kontrolle des Verhaltens bei Fischen.' *Fortschrifte des Zoologie, 22* : 269-309.

Figler, M.H. (1981) 'The hybridisation of ethology and psychopharmacology in the investigation of aggressive behaviour in fish' in P.F. Brain and D. Benton (eds.), *Multidisciplinary approaches to aggression research,* Elsevier/North Holland Press, Amsterdam, pp. 355-67.

Fletcher, D.J., Noe, B.D. and Hunt, E.L. (1978) 'Effects of hexoses on insulin biosynthesis, content and release in channel catfish islets.' *General and Comparative Endocrinology, 35* : 121-6.

Flood, N.B. and Overmeir, J.B. (1981) 'Learning in teleost fish: role of the telencephalon' in P.R. Laming (ed.), *Brain Mechanisms of Behaviour in Lower Vertebrates,* University Press, Cambridge.

Floody, O.R. and Pfaff, D.W. (1974) 'Steroids, hormones and aggressive behaviour.' *Research Publications of the Association for Research in Nervous and Mental Disorders, 52* : 149-85.

Flowerdew, M.W. and Grove, D.J. (1979) 'Some observations of the effects of body weight, meal size and quality on gastric emptying time in the turbot, *Scophthalmus maximus (L.)* using radiography.' *Journal of Fish Biology, 14* : 229-38.

Ford, E. (1921) 'A contribution to our knowledge of the life histories of the dogfishes landed at Plymouth.' *Journal of the Marine Biological Association U.K. 12* : 468-505.

Forselius, S. (1957) 'Studies on anabantid fishes: III.' *Zoologischke Bidrag. Uppsala, 32* : 379-597.

Fostier, A., Weil, C., Terqui, M., Breton, B. and Jalabert, B. (1978) 'Plasma oestradiol-17β and gonadotropin during ovulation in rainbow trout *(Salmo Gairdneri R.).'* *Annales de biologie animale biochimie biophysieque, 18* : 929-36.

Fouchereau-Peron, M., Laburthe, M., Besson, J., Rosselin, G. and Le Gal, Y. (1980) 'Characterization of the vasoactive intestinal polypeptide (VIP) in the gut of fishes.' *Comparative Biochemistry and Physiology, 65A* : 489-92.

Franchi, L.L., Mandl, A.M. and Zuckerman, S. (1962) in S. Zuckerman (ed.), *The Ovary,* Vol. 1, 1-88.

Freadman, M.A. (1981) 'Swimming energetics of striped bass *(Morone saxatilis)* and bluefish *(Pomatomus saltatrix)*: hydrodynamic correlates of locomotion and gill ventilation.' *Journal of Experimental Biology, 90* : 253-65.

Frisch, K. von (1911) 'Beiträge zur Physiologie der Pigmentzellen in der Fisch-haut.' *Pflügers Archiv. 138* : 319-87.

Frost, W.E. (1974) *A survey of the rainbow trout (Salmo gairdneri) in Britain and Ireland,* Salmon and Trout Association, London.

Fryer, G. and Iles, T.D. (1972) *The cichlid fishes of the Great Lakes of Africa,* Oliver & Boyd, Edinburgh.

Fujii, R. and Miyashita, Y. (1976) 'Beta adrenoceptors, cyclic AMP and melanosome dispersion in guppey melanophores' in V. Riley (ed.), *Pigment Cell, 3,* Karger, Basel, 336-44.

Fujii, R. and Miyashita, Y. (1982) 'Receptor mechanisms in fish chromatophores — V. MSH disperses melanosomes in both dermal and epidermal melanophores of a catfish *(Parasilurus asotus).'* *Comparative Biochemistry and Physiology, 71C* : 1-6.

Furness, J.B. and Costa, M. (1980) 'Types of nerves in the enteric nervous system.'

Neuroscience, 5 : 1-20.

Furuichi, M., Nakamura, Y. and Yone, Y. (1980) 'A radioimmunoassay method for determination of fish plasma insulin.' *Bulletin of the Japanese Society of Scientific Fisheries, 46* : 1177-81.

Gall, G.A.E. (1974) 'Influence of size of eggs and age of female on hatchability and growth in rainbow trout.' *California Fish and Game, 60* : 16-35.

Gannon, B.J. and Burnstock, G. (1969) 'Excitatory adrenergic innervation of the fish heart.' *Comparative Biochemistry and Physiology, 29* : 765-74.

Gannon, B.J., Campbell, G. and Satchell, G.H. (1972) 'Monoamine storage in relation to cardiac regulation in the Port-Jackson shark, *Heterodontus portusjacksoni.*' *Zeitschrift für Zellforschung und Mikroskopische Anatomie, 131* : 437-50.

Gerich, J.E., Charles, M.A. and Grodsky, G.M. (1976) 'Regulation of pancreatic insulin and glucagon secretion.' *Annual Review of Physiology, 38* : 353-88.

Gershon, M.D. (1981) 'The Enteric Nervous System.' *Annual Review Neuroscience, 4* : 227-72.

Gibbins, I.L. (1982) 'Lack of correlation between ultrastructure and pharmacological types of non-adrenergic autonomic nerves.' *Cell Tissue Research, 221* : 551-82.

Goetz, F.W. and Theofan, G. (1979) '*In vitro* stimulation of germinal vesicle breakdown and ovulation of yellow perch *(Perca flavescens)* oocytes. Effects of 17α-hydroxy-20β-dihydroprogesterone and prostaglandins.' *General and Comparative Endocrinology, 37* : 273-85.

Goldstein, L. (1982) 'Gill nitrogen excretion' in D.F. Houlihan, J.C. Rankin and T.J. Shuttleworth (eds.), *Gills,* University Press, Cambridge, pp. 193-206.

Gona, O. (1979) 'Toxic effects of mammalian prolactin on *Colisa lalia* and two other related teleostean fish.' *General and Comparative Endocrinology, 37* : 468-73.

Gottfried, H. (1964) 'The occurrence and biological significance of steroids in lower vertebrates. A review.' *Steroids, 3* : 219-42.

Govyrin, V.A. (1977) 'Development of vasomotor adrenergic innervation in onto- and phylogenesis.' *Journal of Evolutionary Biochemistry and Physiology, 13* : 614-20.

Grove, D.J. (1969a) 'The effects of adrenergic drugs on melanophores of the minnow, *Phoxinus phoxinus (L.).*' *Comparative Biochemistry and Physiology, 28* : 37-54.

Grove, D.J. (1969b) 'Melanophore dispersion in the minnow *Phoxinus phoxinus (L.).*' *Comparative Biochemistry and Physiology, 28* : 55-65.

Grove, D.J. and Campbell, G. (1979a) 'The role of extrinsic and intrinsic nerves in the coordination of gut motility in the stomachless flatfish *Rhombosolea tapirina* and *Ammotretis rostrata* Guenther.' *Comparative Biochemistry & Physiology, 63C* : 143-59.

Grove, D.J. and Campbell, G. (1979b) 'Effects of extrinsic nerve stimulation on the stomach of the flathead, *Platycephalus bassensis.*' Cuvier & Valenciennes. *Comparative Biochemistry & Physiology, 63C* : 373-80.

Grove, D.J. and Crawford, C. (1980) 'Correlation between digestion rate and feeding frequency in the stomachless teleost, *Blennius pholis L.*' *Journal of Fish Biology, 16* : 235-47.

Grove, D.J., Lozoides, L. and Nott, J. (1978) 'Satiation amount, frequency of feeding and gastric emptying rate in *Salmo gairdneri.*' *Journal of Fish Biology, 12* : 507-16.

Groves, A.B., Collins, G.B. and Trefetren, G.B. (1968) 'Roles of olfaction and vision in choice of spawning site by homing adult chinook salmon *(Oncorhynchus tshawytscha).*' *Journal of the Fisheries Research Board of*

Canada, 25(5) : 867-76.

Guthrie, D.M. (1981) 'The properties of the visual pathway of a common freshwater fish *(Perca fluviatilis)* in relation to its visual behaviour.' *Society of Experimental Biology,* Seminar Series 9 : 79-112.

Guthrie, D.M. and Banks, J.R. (1974) 'Input characteristics of the intrinsic cells of the optic tectum of teleost fish.' *Comparative Biochemistry and Physiology, 47A,* 83-92.

Guthrie, D.M. and Banks, J.R. (1978) 'The receptive field structure of visual cells from the optic tectum of the freshwater perch.' *Brain Research, 141* : 211-25.

Guraya, S.S. (1978) 'Maturation of the follicular wall of non-mammalian vertebrates' in R.E. Jones (ed.) *The Vertebrate Ovary,* Plenum Press, New York, pp. 261-330.

Guraya, S.S. (1979) 'Recent advances in the morphology, cytochemistry and function of Balbiani's vitelline body in animal oocytes.' *International Review of Cytology, 59* : 249-321.

Gwyther, D. and Grove, D.J. (1981) 'Gastric emptying in *Limanda limanda (L.)* and the return of appetite.' *Journal of Fish Biology, 18* : 245-59.

Hara, T.J., Ueda, J., and Gorbman, A. (1965) 'Influences of thyroxine and sex hormones upon optically evoked potentials in the optic tectum of goldfish.' *General and Comparative Endocrinology, 5* : 313-19.

Hara, T.J. and Gorbman, A. (1967) 'Electrophysiological studies of the olfactory system of the goldfish, I. Modification of the electrical activity of the olfactory bulb by other central nervous structures.' *Comparative Biochemistry and Physiology, 21* : 185-200.

Harris, J.E. (1952) 'A note on the breeding season, sex ratio and embryonic development of the dogfish *Scyliorhinus canicula (L.).' Journal of the Marine Biological Association, U.K., 31* : 269-75.

Hasler, A.D. (1966) *Underwater Guideposts,* University of Wisconsin Press, Madison.

Hasler, A.D. and Wisby, W.J. (1951) 'Discrimination of stream odors by fishes and relation to parent stream behavior.' *American Naturalist, 85* : 223-38.

Hasler, A.D., Scholz, A.T. and Horrall, R.M. (1978) 'Olfactory imprinting and homing in salmon.' *American Scientist, 66(3)* : 347-55.

Hay, T.F. (1978) 'Filial imprinting in the convict cichlid, *Cichlasoma nigrofasciatum.' Behaviour, 65* : 138-54.

Hayashi, S. and Ooshiro, Z. (1975) 'Gluconeogenesis and glycolysis in isolated perfused liver of the eel.' *Bulletin of the Japanese Society of Scientific Fisheries, 41* : 201-8.

Hays, R.O., Levine, S.D., Myers, J.D., Heinemann, H.O., Kaplan, M.A., Franki, N. and Berliner, H. (1976) 'Urea transport in the dogfish kidney.' *Journal of Experimental Zoology, 199* : 309-16.

Hazard, T.P. and Eddy, R.E. (1951) 'Modification of the sexual cycle in brook trout *(Salvelinus fontinalis)* by control of light.' *Transactions of the American Fisheries Society, 80* : 158-62.

Heiligenberg, W. (1974) 'Processes governing behavioural states of readiness.' *Advances in the Study of Behaviour: 5.*

Henderson, I.W. and Wales, N.A.M. (1974) 'Renal diuresis and antidiuresis after injections of arginine vasotocin in the freshwater eel *(Anguilla anguilla L.).' Journal of Endocrinology, 61* : 487-500.

Henderson, I.W. and Brown, J.A. (1980) 'Hormonal actions on single nephron function in teleosts' in B. Lahlou (ed.), *Epithelial transport in the lower vertebrates,* University Press, Cambridge, pp. 163-170.

Henderson, I.W., Brown, J.A., Oliver, J.A. and Haywood, G.P. (1978) 'Hormones and single nephron function in fishes' in P.J. Gaillard and H.H. Boer (eds.),

Comparative Endocrinology, Elsevier/North Holland, Amsterdam, pp. 217-22.

Henderson, I.W., Jotisankasa, V., Moseley, W. and Oguri, M. (1976) 'Endocrine and environmental influences upon plasma cortisol concentrations and plasma renin activity of the eel, *Anguilla anguilla L.' Journal of Endocrinology, 70* : 81-95.

Henderson, N.E. (1963) 'Influence of light and temperature on the reproductive cycle of the eastern brook trout, *Salvelinus fontinalis.' Journal of the Fisheries Research Board of Canada, 20* : 859-97.

Henry, M.G. and Atchison, G.J. (1979) 'Influence of social rank on the behaviour of bluegill, *Lepomis macrochirus* Rafinesque, exposed to sublethal concentrations of cadmium and zinc.' *Journal of Fish Biology, 15* : 309-15.

Hickman, C.P. (1968), 'Urine composition and kidney tubular function in southern flounder, *Paralichthys lethostigma*, in sea water.' *Canadian Journal of Zoology, 46* : 439-55.

Hickman, C.P. and Trump, B.R. (1969) 'The kidney' in W.S. Hoar and D.J. Randall (eds.), *Fish Physiology Vol. 1, Excretion, Ionic Regulation and Metabolism*, Academic Press, New York, pp. 91-239.

Hinde, R.A. (1970), *Animal Behaviour*, McGraw Hill, New York.

Hinde, R.A. (1982), *Ethology*, Fontana, Glasgow.

Hirano, T. and Mayer-Gostan, N. (1978) 'Endocrine control of osmoregulation in fish' in P.J. Gaillard and H.H. Boer, (eds.), *Comparative Endocrinology*, Elsevier/North Holland, Amsterdam, pp. 209-12.

Hirsch, P.J. (1977) 'Conditioning of the heart rate of coho salmon *(Oncorhynchus kisutch)* to odors.' Unpublished Ph.D. Thesis, University of Wisconsin, Madison, U.S.A.

Hisaw, F.L. and Albert, A. (1947) 'Observations on the reproduction of the spiny dogfish *(Squalus acanthias).' Biological Bulletin of Marine Biology Laboratory, Woods Hole, 92* : 187-99.

Hisaw, F.L. Jr. and Hisaw, F.L. (1959) 'The corpora lutea of elasmobranch fishes.' *Anatomical Record, 135* :269-78.

Hiyama, Y., Taniuchi, T., Suyama, K., Ishoka, K., Sato, R., Kajihara, T. and Maiwa, R. (1966) 'A preliminary experiment on the return of tagged chum salmon to the Otsucki River, Japan.' *Japan Society Scientific Fisheries, 33* : 18-19.

Hoar, W.S. (1962) 'Hormones and the reproductive behaviour of the male three-spined stickleback *(Gasterosterus aculeatus).' Animal Behaviour, 10* : 247-66.

Hoar, W.S. (1969) 'Reproduction' in W.S. Hoar and D.J.Randall (eds.), *Fish Physiology, Vol. 3*, Academic Press, New York.

Hoar, W.S., Randall, D.J. and Brett, J.R. (1979), *Fish Physiology, Vol. VIII.* Academic Press, New York.

Holmes, H.L. and Ball, R.N. (1974), *The Pituitary Gland*, University Press, Cambridge, U.K.

Holmgren, S. (1977) 'Regulation of the heart of a teleost, *Gadus morhua*, by autonomic nerves and circulating catecholamines.' *Acta Physiologica Scandinavica, 99* : 62-74.

Holmgren, S. (1978) 'Sympathetic innervation of the coeliac artery from a teleost, *Gadus morhua.' Comparative Biochemistry and Physiology, 60C* : 27-32.

Holmgren, S. (1982) 'The effects of VIP, substance P and ATP on isolated muscle strips from the stomach of the rainbow trout, *Salmo gairdneri.' Acta Physiologica Scandinavica, 114* : 39A.

Holmgren, S. and Nilsson, S. (1974) 'Drug effects on isolated artery strips from two teleosts, *Gadus morhua* and *Salmo gairdneri.' Acta Physiologica Scandinavica, 90* : 431-7.

Holmgren, S., and Nilsson, S. (1975) 'Effects of some adrenergic and cholinergic drugs on isolated spleen strips from the cod, *Gadus morhua.' European Journal of Pharmacology, 32* : 163-9.

Holmgren, S. and Nilsson, S. (1976) 'Effects of denervation, 6-hydroxydopamine and reserpine on the cholinergic and adrenergic responses of the spleen of the cod, *Gadus morhua.' European Journal of Pharmacology, 39* : 53-9.

Holmgren, S. and Nilsson, S. (1981) 'On the non-adrenergic, non-cholinergic innervation of the rainbow trout stomach.' *Comparative Biochemistry and Physiology, 70C* : 65-9.

Holmgren, S. and Nilsson, S. (1982) 'Minireview. Neuropharmacology of adrenergic neurons in teleost fish.' *Comparative Biochemistry and Physiology* : (In press).

Holmgren, S., Vaillant, C., and Dimaline, R. (1982) 'VIP-, substance P-, gastrin/CCK-, bombesin-, somatostatin-, and glucagon-like immunoreactivities in the gut of the rainbow trout, *Salmo gairdneri.' Cell and Tissue Research, 223* : 141-53.

Holmqvist, A.L., Dockray, G.J., Rosenquist, G.L. and Walsh, J.G. (1979) 'Immunochemical characterization of cholecystokinin-like peptides in lamprey gut and brain.' *General and Comparative Endocrinology, 37* : 474-81.

Hoover, E.E. and Hubbard, H.E. (1937) 'Modification of the sexual cycle in trout by control of light.' *Copeia, 1937 (4)* : 206-10.

Htun-Han, M. (1975) 'The Effects of Photoperiod on Maturation in the Dab, *Limanda limanda*, unpublished Ph.D. Thesis, University of East Anglia, U.K.

Hughes, G.M. (1972) 'The relationship between cardiac and respiratory rhythms in the dogfish, *Scyliorhinus canicula L.' Journal of Experimental Biology, 57* : 415-34.

Hughes, G.M. and Morgan, M. (1973) 'The structure of fish gills in relation to their respiratory function.' *Biological Reviews, 48* : 419-75.

Hughes, G.M., Peyraud, C., Peyraud-Waitzenegger, M., and Soulier, P. (1981) 'Proportion of cardiac output concerned with gas exchange in the gills of the eel *(A. anguilla).' Journal of Physiology, London. 310* : 61-62P.

Hunter, G.A., Donaldson, E.M. and Dye, H.M. (1981) 'Induced ovulation in coho salmon *(O. kisutch).* 1. Further studies on the use of salmon pituitary preparations.' *Aquaculture, 26* : 117-27.

Huntingford, F.A. (1976) 'A comparison of the reaction of sticklebacks in different reproductive conditions towards conspecifics and predators.' *Animal Behaviour, 24* : 694-7.

Huntingford, F.A. (1979) 'Pre-breeding aggression in male and female three-spined sticklebacks *(Gasterosteus aculeatus).' Aggressive Behaviour, 5* : 51-8.

Huntingford, F.A. (1982) 'Do inter- and intraspecific aggression vary in relation to predation pressure in sticklebacks?' *Animal Behaviour, 30* : 909-16.

Hurk, R. van den Peute, J. (1979) 'Cyclic changes in the ovary of the rainbow trout, *Salmo gairdneri* with special reference to sites of steroidogenesis.' *Cell and Tissue Research, 199*: 289-306.

Hutchison, J.B. (1978) *Biological Determinants of Sexual Behavior,* John Wiley and Sons, New York.

Idler, D.R., Bazar, L.S. abd Hwang, S.J. (1975) 'Fish gonadotropin(s): Isolation of gonadotropin(s) from chum salmon pituitary glands using affinity chromatography.' *Endocrine Research Communications, 2* : 215-35.

Idler, D.R., Bitners, I.I. and Schmidt, P.J. (1961) '11-ketotestosterone: an androgen for sockeye salmon.' *Canadian Journal of Biochemistry and Physiology, 39* : 1737-42.

Idler, D.R., McBride, J.R., Jones, R.E.E. and Tomlinson, N. (1961) 'Olfactory perception in migrating salmon, II: studies on a laboratory bioassay for a homestreamwater and mammalian repellent.' *Canadian Journal of Biochemistry & Physiology, 39* : 1575-84.

Ince, B.W. (1979) 'Insulin secretion from the *in situ* perfused pancreas of the European silver eel *(Anguilla anguilla L.).*' *General and Comparative Endocrinology, 37* : 533-40.

Ince, B.W. (1980) 'Amino acid stimulation of insulin secretion from the perfused eel pancreas : modification by somatostatin, adrenalin, and theophylline.' *General and Comparative Endocrinology, 40* : 275-82.

Ince, B.W. (1982) 'Plasma clearance kinetics of unlabelled bovine insulin in rainbow trout *(Salmo gairdneri).*' *General and Comparative Endocrinology, 46* : 463-72.

Ince, B.W. (1983) 'Secretin-stimulated insulin release *in vivo* in European eels, *Anguilla anguilla L.*' *Journal of Fish Biology, 22* : 259-63.

Ince, B.W. and Thorpe, A. (1974) 'Effects of insulin and of metabolite loading on blood metabolites in the European silver eel *(Anguilla anguilla L.).*' *General and Comparative Endocrinology, 23* : 460-71.

Ince, B.W. and Thorpe, A. (1975) 'Hormonal and metabolite effects on plasma free fatty acids in the Northern pike, *Esox lucius L.*' *General and Comparative Endocrinology, 27* : 144-52.

Ince, B.W. and Thorpe, A. (1976) 'The '*in vivo*' metabolism of ^{14}C glucose and ^{14}C-glycine in insulin-treated Northern pike *(Esox lucius L.).*' *General and Comparative Endocrinology, 28* : 481-6.

Ince, B.W. and Thorpe, A. (1977a) 'Plasma insulin and glucose responses to glucagon and catecholamines in the European silver eel *(Anguilla anguilla L.).*' *General and Comparative Endocrinology, 33* : 453-9.

Ince, B.W. and Thorpe, A. (1977b) 'Glucose and amino acid-stimulated insulin release *in vivo* in the European silver eel *(Anguilla anguilla L.).*' *General and Comparative Endocrinology, 31* : 249-56.

Ince, B.W. and Thorpe, A. (1978a) 'Effects of insulin on plasma amino acid levels in the Northern pike *(Esox lucius L.).*' *Journal of Fish Biology, 12* : 503-6.

Ince, B.W. and Thorpe, A. (1978b) 'Insulin kinetics and distribution in rainbow trout *(Salmo gairdneri).*' *General & Comparative Endocrinology, 35* : 1-9.

Inui, Y. and Yokote, M. (1977) 'Effects of glucagon on amino acid metabolism in Japanese eels *(Anguilla japonica).*' *General and Comparative Endocrinology, 33* : 167-73.

Inui, Y., Arai, S. and Yokote, M. (1975) 'Gluconeogenesis in the eel. VI. Effects of hepatectomy, alloxan and mammalian insulin on the behaviour of plasma amino acids.' *Bulletin of the Japanese Society of Scientific Fisheries, 41* : 1105-11.

Irving, L.D., Solandt, O.Y. and Solandt, D.M. (1935) 'Nerve impulses from branchial pressure receptors in the dogfish.' *Journal of Physiology, London, 84* : 187-90.

Ito, H. (1970) 'Fine structures of the carp optic tectum.' *Journal für Hirnforschung, 12* : 325-54.

Iwamatsu, T. and Keino, H. (1978) 'Scanning electron microscopic study on the surface change of eggs of the teleost, *Oryzias latipes*, at the time of fertilization.' *Development, Growth and Differentiation, 20* : 237-50.

Jackim, E. and La Roche, G. (1973) 'Protein synthesis in '*Fundulus heteroclitus*' muscle.' *Comparative Biochemistry and Physiology, 44A* : 851-6.

Jalabert, B. (1976) '*In vitro* oocyte maturation and ovulation in rainbow trout *(Salmo gairdneri)* northern pike *(Esox lucius)* and goldfish *(Carassius auratus).*' *Journal of Fisheries Research Board of Canada, 33* : 974-88.

Jalabert, B. and Breton, B. (1980) 'Évolution de la gonadotropine plasmatique t-GTH aprés l'ovulation chez la truite arc-en-ciel *(Salmo gairdneri)* et influence de la rétention des ovules.' *Comptes Rendus Hebdomadaires de Séances de l'Académie des Sciences Series D, 290* : 799-801.

Jalabert, B., Breton, B. and Bry, C. (1980). 'Évolution de la gonadotropine plasmatique t-GTH après synchronisation des ovulations par injection de 17α-hydroxy-20β-dihydroprogestérone chez la truite arc-en-ciel *(Salmo gairdneri R.).*' *Comptes Rendus Hebdomadaires de Séances de l'Académie des Sciences Series D, 290* : 1431-4.

Janzen, W. (1933) 'Untersuchunge über Grosshirn funktionen des Goldfisches *(Carassius auratus).*' *Zoologischer Jahrbuch, 52* : 591-628.

Jenkin, P. (1928) 'Note on the sympathetic nervous system of *Lepidosiren paradoxa.*' *Proceedings of the Royal Society of Edinburgh, 48* : 55-69.

Jenkins, N. (1978) 'The endocrine control of reproduction in the dogfish *(Scyliorhinus canicula L.),* with special reference to the hypothalamus.' Unpublished Ph.D. thesis, University of Wales, U.K.

Jenkins, N. and Dodd, J.M. (1980) 'Effects of synthetic mammalian gonadotrophin releasing hormone and dogfish hypothalamic extracts on levels of androgens and oestradiol in the circulation of the dogfish *(Scyliorhinus canicula L.).*' *Journal of Endocrinology, 86* : 171-7.

Jenkins, N., Joss, J.P. and Dodd, J.M. (1980) 'Biochemical and autoradiographic studies on the oestradiol-concentrating cells in the diencephalon and pituitary gland of the female dogfish *(Scyliorhinus canicula L.).*' *General and Comparative Endocrinology, 40* : 211-19.

Jenkins, T.M. (1969) 'Social structure, position choice, and microdistribution of two trout species *(Salmo trutta* and *S. gairdneri)* resident in mountain streams.' *Animal Behaviour Monographs, 2* : 57-125.

Jensen, A. and Duncan, R. (1971) 'Homing in transplanted coho salmon.' *Progressive Fish Culturalist, 33* : 216-18.

Jerison, H.J. (1973) *Evolution of the brain and intelligence,* Academic Press, New York.

Jobling, M. (1980) 'Gastric evacuation in plaice, *Pleuronectes platessa L.* Effect of dietary energy level and food composition.' *Journal of Fish Biology, 17* : 187-96.

Jobling, M. (1981) 'Mathematical models of gastric emptying and the estimation of daily rates of food consumption for fish.' *Journal of Fish Biology, 19* : 245-57.

Jobling, M., Gwyther, D. and Grove, D. (1977) 'Some effects of temperature, meal size and body weight on gastric evacuation time in the dab, *Limanda limanda (L.).*' *Journal of Fish Biology, 10* : 291-8.

Johansen, K. and Burggren, W. (1980) 'Cardiovascular function in the lower vertebrates' in G.H. Bourne (ed.), *Heart and heart-like organs, Vol. 1,* Academic Press, New York, pp. 61-117.

Johnels, A.G. (1956) 'On the peripheral autonomic nervous system of the trunk region of *Lampetra planeri.*' *Acta Zoologica, 37* : 251-86.

Johns, L.S. and Liley, N.R. (1970) 'The effects of gonadectomy and testosterone treatment on the reproductive behaviour of the male blue gourami, *Trichogaster trichopterus.*' *Canadian Journal of Zoology, 48* : 977-87.

Jones, D.R. and Milson, W.K. (1982) 'Peripheral receptors affecting breathing and cardiovascular function in non-mammalian vertebrates.' *Journal of Experimental Biology, 100* : 59-91.

Jones, D.R. and Randall, D.J. (1978) 'The respiratory and circulatory systems during exercise' in W.S. Hoar and D.J. Randall (eds.), *Fish Physiology, Vol. VII,* Academic Press, New York, pp. 425-501.

Jones, D.R., Randall, D.J. and Jarman, G.M. (1970) 'A graphical analysis of oxygen transfer in fish.' *Respiration Physiology, 10* : 285-98.

268   *References*

Jones, G.P. (1983) 'Relationship between density and behaviour in juvenile *Pseudolabrus celidotus* (Pisces:Labridae).' *Animal Behaviour.* (In press).

Kagawa, H., Takano, K. and Nagahama, Y. (1981) 'Correlation of plasma estradiol-17β and progesterone levels with ultrastructure and histochemistry of ovarian follicles in the white-spotted char, *Salvelinus leucomacnis.*' *Cell and Tissue Research, 218* : 315-29.

Kanatani, H. and Nagahama, Y. (1980) 'Mediators of oocyte maturation.' *Biomedical Research, 1* :273-91.

Kapoor, B.G., Smit, H. and Verighina, I.A. (1975) 'The alimentary canal and digestion in teleosts.' *Advances in Marine Biology, 13* : 109-239.

Kato, T. (1975) 'The relation between growth and reproductive characters of rainbow trout, *Salmo gairdneri.*' *Bulletin of the Freshwater Fisheries Research Laboratory, Tokyo, 25* : 83-115.

Katzir, G. (1981) 'Visual aspects of species recognition in the damselfish, *Dascyllus aruanus.*' *Animal Behaviour, 29(3)* : 842.

Kawajiri, M. (1927) 'On the optimum temperature of water for hatching the eggs of rainbow trout.' *Journal of the Imperial Fisheries Institute of Tokyo, 23* : 59-65.

Kelly, M.J., Moss, R.L. and Dudley, C.A. (1976) 'Differential sensitivity of preoptic-septal neurones to microelectrophoresed oestrogen during the oestrous cycle.' *Brain Research, 114* : 152-7.

Kennedy, G.C. (1953) 'The role of depot fat in the hypothalamic control of food intake in the cat.' *Proceedings of the Royal Society, B140* : 578-92.

Kim, Y.S., Stumpf, W.E., Sar, M. and Martinez-Vargas, M.C. (1978) 'Oestrogen and androgen target cells in the brain of fishes, reptiles and birds : Phylogeny and ontogeny.' *American Zoologist, 18* : 425-33.

Kime, D.E. (1978) 'Steroid biosynthesis by the testis of the dogfish *Scyliorhinus caniculus.*' *General and Comparative Endocrinology, 34* : 6-17.

Kime, D.E. (1980) 'Comparative aspects of testicular androgen biosynthesis in non-mammalian vertebrates' in G. Delrio and J. Bracket (eds.), *Steroids and Their Mechanism of Action in Non-mammalian Vertebrates,* Raven Press, New York.

Kimmel, C.B., Eaton, R.C. and Powell, S.L. (1980) 'Decreased fast-start performance of zebra-fish lacking Mauthner neurons.' *Journal of Comparative Physiology, 140* : 343-50.

King, J.A. and Millar, R.P. (1979) 'Heterogeneity of vertebrate luteinising hormone-releasing hormone.' *Science, 206* : 67-9.

King, J.A. and Millar, R.P. (1980) 'Comparative aspects of luteinising hormone-releasing hormone structure and function in vertebrate phylogeny.' *Endocrinology, 106* : 707-17.

Kirby, S. and Burnstock, G. (1969) 'Comparative pharmacological studies of isolated spiral strips of large arteries from lower vertebrates.' *Comparative Biochemistry and Physiology, 28* : 307-19.

Kitabgi, P. and Vincent, J.P. (1981) 'Neurotensin is a potent inhibitor of guinea-pig colon contractile activity.' *European Journal of Pharmacology, 74* : 311-18.

Knowles, F., Vollrath, L. and Meurling, P. (1975) 'Cytology and neuro-endocrine relations of the pituitary of the dogfish, *Scyliorhinus canicula.*' *Proceedings of the Royal Society, B, 191* : 507-25.

Kobayashi, K.A. and Wood, C.M. (1980) 'The response of the kidney of the freshwater rainbow trout to true metabolic acidosis.' *Journal of Experimental Biology, 84* : 227-44.

Kodric-Brown, A. (1978) 'Establishment and defence of breeding territories in a pupfish *(Cyprinodontidae : Cyprinodon).*' *Animal Behaviour 26* : 818-34.

Konishi, J. (1960) 'Electrical response of the visual center in fish, especially to coloured light flash.' *Japanese Journal of Physiology, 10* : 13-17.

Korn, H. and Faber, D.S. (1978) *Neurobiology of the Mauthner Cell,* Raven Press, New York.

Kunesh, W.H., Freshman, W.J., Hoehm, M. and Nordin, N.G. (1974) 'Altering the spawning cycle of rainbow trout by control of artificial light.' *Progressive Fish Culturist, 36* : 225-6.

Kuzimina, V.V. (1966) 'The influence of hormonal factors on the feeding reaction in fish.' *Dokl. Akad. Nauk. SSSR. 170* : 486-8.

Kyle, A.L. and Peter, R.E. (1979) 'Effect of brain lesions on spawning behaviour in the male goldfish.' *Canada West Society for Reproductive Biology Meeting,* University of Saskatchewan.

Laale, H.W. (1980) 'The perivitelline space and egg envelopes of bony fishes: a review.' *Copeia 1980* : 210-26.

Lahlou, B.(1966) 'Mise en évidence d'un "recrutément glomerulaire" dans le rein des Téléostéens d'après la méthode du Tm glucose.' *Comptes Rendues de l'Académie des Sciences, Paris, 262* : 1356-8.

Lahlou, B., Henderson, I.W. and Sawyer, W.H. (1969) 'Renal adaptations by *Ospanus tau,* a euryhaline aglomerular teleost, to dilute media.' *American Journal of Physiology, 216* : 1266-72.

Laird, L.M., Wilson, A.R. and Holliday, F.G.T. (1981) 'Field trials of a method of induction of autoimmune gonad rejection in Atlantic salmon *(Salmo salar L.).' Reproduction Nutrition Développement, 20* : 1781-88

Lam, T.J. (1982) 'Applications of endocrinology to fish culture.' *Canadian Journal of Fisheries and Aquatic Sciences, 39* : 111-137.

Lam, T.J., Nagahama, Y., Chan, K and Hoar, W.S. (1978) 'Overripe eggs and postovulatory corpora lutea in the three-spined stickleback, *Gasterosteus aculeatus L., form trachurus.' Canadian Journal of Zoology, 56* : 2029-36.

Lambert, J.G.D. and Bohemen, Ch.G. van (1979) 'Steroidogenesis in the ovary of the rainbow trout, *Salmo gairdneri* during the reproductive cycle.' *Proceedings of the Indian Academy of Sciences B, 45* : 414-20.

Lambert, J.G.D. and Bohemen, Ch.G. van (1980) 'Oestrogen synthesis in the female trout *Salmo gairdneri.' General and Comparative Endocrinology, 40* : 323-4

Lambert, J.G.D. and van Oordt, P.G.W.J. (1974) 'Ovarian hormones in teleosts.' *Fortschritte der Zoologie, 22* : 340-9.

Lambert, J.G.D., Bosman, G.I.C.C.M., Hurk, R. van den and Oordt, P.G.W.J. van (1978) 'Annual cycle of plasma oestradiol-17ß in the female trout, *Salmo gairdneri.' Annales de biologie animale biochimie biophysique, 18* : 923-7.

Laming, P.R. (1981) 'The physiological basis of alert behaviour in fish' *Brain mechanisms of behaviour in lower vertebrates. Society for Experimental Biology,* Seminar Series *9* : 203-24.

Lance, V. and Callard, I.P. (1969) 'A histochemical study of ovarian function in the ovoviviparous elasmobranch, *Squalus acanthias.' General and Comparative Endocrinology, 13* : 255-67.

Lance, V. and Callard, I.P. (1978) 'Gonadotrophic activity in pituitary extracts from an elasmobranch *(Squalus acanthias L.).' Journal of Endocrinology, 78* : 149-50.

Langer, M., Van Noorden, S., Polak, J.M. and Pearse, A.G.E. (1979) 'Peptide hormone-like immunoreactivity in the gastrointestinal tract and endocrine pancreas of eleven teleost species.' *Cell and Tissue Research, 199* : 493-508.

Langley, J.N. (1921) *The autonomic nervous system,* Cambridge University Press, U.K.

Larsson, A and Lewander, K. (1972) 'Effects of glucagon administration to eels

(Anguilla anguilla L.).' Comparative Biochemistry and Physiology, 43A : 831-6.

Larsson, L-I. (1980) 'Gastrointestinal cells producing endocrine, neurocrine and paracrine messengers.' *Clinics in Gastroenterology, 9* : 485-516.

Larsson L-I. and Rehfeld, J.F. (1978) 'Evolution of CCK-like hormones' in S.R. Bloom (ed.) *Gut Hormones,* Churchill-Livingstone, Edinburgh, pp. 68-73.

Laurent, P. (1962) 'Contribution à l'étude morphologique et physiologique de l'innervation du coeur des téléostéens.' *Archives d'Anatomie Microscopique et de Morphologie Experimentale, 51* : 339-458.

Laurent, P. (1967) 'Neurophysiologie — la pseudobranchie des téléostéens: preuves electrophysiologiques de ses fonctions chémoréceptrices et baroréceptrices.' *Comptes Rendus Hebdomadaire des Séances de l'Académie des Sciences, Paris, 264* : 1879-82.

Leibson, L.G. (1972) 'Some peculiarities of metabolism and its endocrine regulation in fishes of different motor activity.' *Journal of Evolutionary Biochemistry and Physiology, 8* : 280-8. (In Russian).

Leshner, A.I. (1978) *An introduction to behavioural endocrinology,* University Press, Oxford.

Leshner, A.I. (1981) 'The role of hormones in the control of submissiveness' in P.F. Brain & D. Benton (eds.) *Multidisciplinary approaches to aggression research,* Elsevier/North Holland, Amsterdam, pp 309-31.

Lewander, K., Dave, G., Johansson-Sjobeck, M.L., Larsson, A. and Lidman, U. (1976) 'Metabolic effects of insulin in the European eel *(Anguilla anguilla).' General and Comparative Endocrinology, 29* : 455-67.

Lewis, M.R. (1980) 'Rainbow trout: production and marketing.' *University of Reading, Department of Agricultural Economics and Management Miscellaneous Study,* No. 68.

Lewis, M. and Dodd, J.M. (1974) 'Thyroid function and the ovary in the spotted dogfish *Scyliorhinus canicula.' Journal of Endocrinology, 63* : 63.

Leydig, F. (1853) *Anatomische-histologische Untersuchnungen über Fische und Reptilen,* Reimer, Berlin.

Liley, N.R. (1969) 'Hormones and reproductive behaviour in fishes' in W.S. Hoar & D.J. Randall (eds.) *Fish Physiology, Vol. III,* Academic Press, New York, pp 73-160.

Liley, N.R. (1972) 'The effects of estrogens and other steroids on the sexual behaviour of the female guppy, *Poecilia reticulata.' General and Comparative Endocrinology, Supplement 3* : 542-52.

Liley, N.R. (1980) 'Patterns of hormonal control in the reproductive behaviour of fish and their relevance to fish management and culture programs' in J.E. Bardach (ed.) *The Physiological and Behavioral Manipulation of Food Fish as Production and Management Tools,* ICLARM, Manila. (In press).

Liley, N.R. and Donaldson, E.M. (1969) 'The effects of fish pituitary material on the behaviour of hypophysectomized female guppies, *Poecilia reticulata* Peters.' *Canadian Journal of Zoology, 47* : 569-73.

Liley, N.R. and Wishlow, W. (1974) 'The interaction of endocrine and experiential factors in the regulation of sexual behaviour in the female guppy *Poecilia reticulata.' Behaviour, 48* : 185-214.

Lincoln, R.F. (1981a) 'Sexual maturation in triploid male plaice *(Pleuronectes platessa)* and plaice × flounder *(Platichtys flesus)* hybrids.' *Journal of Fish Biology, 19* : 415-26.

Lincoln, R.F. (1981b) 'Sexual maturation in female triploid plaice *(Pleuronectes platessa)* and plaice × flounder *(Platichtys flesus)* hybrids.' *Journal of Fish Biology, 19* : 499-507.

References 271

Lincoln, R.F. and Scott, A.P. (1983) 'Production of all-female triploid rainbow trout.' *Aquaculture, 30* : 375-80.

Lincoln, R.F., Aulstad, D. and Grammeltveldt, A. (1974) 'Attempted triploid induction in Atlantic salmon *(Salmo salar)* using cold shocks.' *Aquaculture, 4* : 287-97.

Loewi, O. and Navratil, E. (1926) 'Über humorale Übertragbarkeit der Herznervenwirkung. I. Mitteilung.' *Pflügers Archiv. 189* : 239-42.

Logan, A.G., Moriarty, R.J., Morris, R and Rankin, J.C. (1980) 'The anatomy and blood system of the kidney in the river lamprey, *Lampetra fluviatilis.*' *Anatomy and Embryology, 158* : 245-52.

Logan, A.G., Moriarty, R.J. and Rankin, J.C. (1980) 'A micropuncture study of kidney function in the river lamprey, *Lampetra fluviatilis*, adapted to freshwater.' *Journal of Experimental Biology, 85* : 137-47.

Logan, A.G., Morris, R and Rankin, J.C. (1980) 'A micropuncture study of kidney function in the river lamprey, *Lampetra fluviatilis*, adapted to sea water.' *Journal of Experimental Biology, 88* : 239-47.

Lomholt, J.P. and Johansen, K. (1979) 'Hypoxia acclimation in carp — how it affects oxygen uptake, ventilation and O_2 extraction from water.' *Physiological Zoology, 52* : 38-49.

Lovell, R.T. (1979) 'Factors affecting voluntary food consumption of the channel catfish.' *Proceedings Annual Conference South East Association of Fish & Wildlife Agencies, 33* : 563-71. *(American Society of Fisheries 11 (No. 11)).*

Ludwig, B., Higgs, D.A., Fagerlund, U.H.M. and McBride, J.R. (1977) 'A preliminary study of insulin participation in the growth regulation of coho salmon *(Oncorhynchus kisutch).*' *Canadian Journal of Zoology, 55* : 1756-8.

Lukomskaya, N.J. and Michelson, M.J. (1972) 'Pharmacology of the isolated heart of the lamprey, *Lampetra fluviatilis*,' *Comparative and General Pharmacology, 3* : 213-25.

Macey, M.J., Pickford, G.E. and Peter, R.E. (1974) 'Forebrain localization of the spawning reflex response to exogenous neurohypophyseal hormones in the killifish, *Fundulus heteroclitus.*' *Journal of Experimental Zoology, 190* : 269-80.

Madison, D.M., Scholz, A.T., Cooper, J.C., Horrall, R.M., Hasler, A.D. and Dizon, A.E. (1973) 'Olfactory hypothesis and salmon migrations: a synopsis of recent findings.' *Fisheries Research Board Canada Technical Report, No. 414* : 1-35.

Magnusson, J.J. (1962) 'An analysis of aggressive behaviour, growth and competition for food and space in medaka *(Oryzias latipes*, Pisces, Cyprinodontidae).' *Canadian Journal of Zoology, 40* : 313-63.

Magnusson, J.J. (1969) 'Digestion and food consumed by skipjack tuna, *Katsuwonus pelamis.*' *Transactions of the American Fisheries Society, 98* : 379-92.

Marza V.D. (1938) *Histophysiologie de l'ovogénèse*, Hermann, Paris.

Matthews, L.H. (1950) 'Reproduction in the basking shark, *Cetorhinus maximus* (Gunner).' *Philosophical Transactions of the Royal Society London, 234B* : 247-316.

Matthews, L.H. (1955) 'The evolution of viviparity in vertebrates.' *Memoirs of the Society of Endocrinology, 4* : 129-48.

Matty, A.J. (1960) 'Thyroid cycles in fish.' *Symposium of the Zoological Society London, 2* : 1-15.

Mayer, J. (1955) 'Regulation of energy intake and the body weight; the glucostatic theory and the lipostatic hypothesis.' *Annals of the N.Y. Academy of Sciences,*

63: 15-42.

Maynard-Smith, J. (1977) 'Parental investment : a prospective analysis' *Animal Behaviour, 25*: 1-9.

Meek, J. and Schellart, N.A.M. (1978) 'A Golgi study of the goldfish optic tectum.' *Journal of Comparative Neurology, 182*: 89-122.

Mellinger, J.C.A. (1960) 'Contribution a l'étude de la vascularisation et du developpement de la région hypophysaire d'un Sélacien, *Scyliorhinus caniculus* L.' *Bulletin Société Zoologique de France 85*: 123-89.

Mellinger, J. (1961) 'La circulation sanguine dans le complexe hypophysaire de la rousette.' *Bulletin Société Zoologique de France, 85*: 395-9.

Mellinger, J. (1963) 'Etude histophysiologique du système hypothalamo-hypophysaire de *Scyliorhinus caniculus* (L.) en état de mélano-dispersion permanente.' *General and Comparative Endocrinology, 3*: 26-45.

Mellinger, J.C.A. (1964) 'Les relations neuro-vasculo-glandulaires dans l'appareil hypophysaire de la roussette, *Scyliorhinus caniculus (L.)* (Poissons Elasmobranchs).' *Archs. Anat. Histol. Embryol, 47*: 1-201.

Mellinger, J.C.A. (1965) 'Stades de la spermatogenèse chez *Scyliorhinus caniculus* (L.): Description, données histochimiques, variations normales et expérimentales.' *Z. Zellforsch. Mikrosk. Anat., 67*: 653-73.

Mellinger, J. (1966a) 'Etude biométrique et histophysiologique des relations entre les gonades, le foie et la thyroide chez *Scyliorhinus caniculus* (L.). Contribution a l'étude des caractères sexuels secondaires des chondrichthyens.' *Cahiers de Biologie Marine VII*: 107-37.

Mellinger, J.C.A. (1966b) 'Variations de la structure hypophysaire chez les chondrichthyens : étude de l'ange de mer (Squatina) et de la pastenague (Trygon).' *Ann. Endocr. 27*: 439-50.

Mellinger, J.C.A. and Dubois, M.P. (1973) 'Confirmation, par l'immuno-fluorescence, de la fonction corticotrope du lobe rostral et de la fonction gonadotrope du lobe ventral de l'hypophyse d'un possion cartilagineux, la torpille marbrée, *(Torpedo marmorata).*' *Comptes Rendus des Scéances Academie Sciences, 276*: 1879-81.

Mellinkoff, S.M., Frankland, M., Boyle, D. and Greipel (1956) 'Relationship between serum amino acid concentration and fluctuations in appetite.' *Journal of Applied Physiology, 8*: 535-8.

Metcalfe, J.D. and Butler, P.J. (1982) 'Differences between directly measured and calculated values for cardiac output in the dogfish : a criticism of the Fick method.' *Journal of Experimental Biology, 99*: 255-68.

Metten, H. (1939) 'Studies on the reproduction of the dogfish.' *Philosophical Transactions of the Royal Society B. 230*: 217-38.

Metuzals, J., Ballintijn-de Vries, G. and Baerands, G.P. (1968) 'The correlation of histological changes in the adenohypophysis of the cichlid fish *Aequidens portalegrensis* (Hensel) with behaviour changes during the reproductive cycle.' *Proc. Kon. Ned. Akad. v. Wet. Ser. C. 71*: 391-410.

Meurling, P. (1960) 'Presence of a portal system in elasmobranchs.' *Nature, London, 187*: 336-7.

Meurling, P. (1967) 'The vascularisation of the pituitary in elasmobranchs.' *Sarsia, 28*: 1-104.

Miles, F.A. (1972) 'Centrifugal control of the avian retina. IV. Effects of reversible cold block of the isthmo-optic tract on the receptive field properties of cells in the retina and isthmo-optic nucleus.' *Brain Research, 48*: 131-45.

Miller, R.J. (1978) 'Agonistic behaviour in fishes and terrestrial vertebrates' in E.S. Reese and F.J. Lighter (eds.), *Contrasts in Behaviour. Adaptations in the aquatic and terrestrial environments*, Wiley & Sons, New York.

Minick, M.C. and Chavin, W. (1972) 'Effects of vertebrate insulins upon serum

FFA and phospholipid levels in the goldfish *Carassius auratus L.' Comparative Biochemistry and Physiology, 41* : 791-804.

Montalambert, G. de, Bry, C. and Billard, C. (1978) 'Control of reproduction in Northern Pike.' *Special Publication of the American Fisheries Society, 11* : 217-25.

Morgan, M. and Tovell, P.W.A. (1973) 'The structure of the gill of the trout,*Salmo gairdneri* (Richardson).' *Zeitschrift für Zellforschung und Mikroskopische Anatomie, 142* : 147-62.

Moriarty, R.J., Logan, A.G. and Rankin, J.C. (1978) 'Measurement of single nephron filtration rate in the kidney of the river lamprey, *Lampetra fluviatilis L.' Journal of Experimental Biology, 77* : 57-69.

Morrell, J.I. and Pfaff, D.W. (1978) 'A neuroendocrine approach to brain function. Localisation of sex steroid concentrating cells in vertebrate brains.' *American zoologist, 18* : 447-60.

Morris, R. (1965) 'Studies on salt and water balance in *Myxine glutinosa (L.).' Journal of Experimental Biology, 42* : 359-71.

Morris, R. (1972) 'Osmoregulation' in M.W. Hardisty and I.C. Potter (eds.), *The Biology of Lampreys; vol. 2*, Academic Press, London, pp. 193-239.

Mott, J.C. (1951) 'Some factors affecting the blood circulation in the common eel *(Anguilla anguilla).'* Journal of Physiology, 114 : 387-98.

Munro, A.D. (1982) *Comparative studies of the forebrain of Aequidens pulcher (Teleostei : Cichlidae), with particular reference to the control of the pituitary gland.* Unpublished PhD Thesis, University of Wales, U.K.

Munro, A.D. & Dodd, J.M. (1983) 'The forebrain of fishes with particular reference to neuroendocrine control mechanisms.' in L. Bolis and G. Nistico (eds.) *Progress in non-mammalian brain research. Volume III.* CRC Press, New York, (In press).

Munro, A.D. and Pitcher, T.J. (1983) 'Steroids and agonistic behaviour in a teleost, *Aequidens pulcher* (Cichlidae).' *Hormones & Behaviour*, submitted.

Murat, J.C., Plisetskaya, E.M. and Woo, N.Y.S. (1981) 'Endocrine control of nutrition in cyclostomes and fish.' *Comparative Biochemistry and Physiology, 68A* : 149-58.

Murat, J.C. and Serfaty, A. (1975) 'Effect of temperature on carbohydrate metabolism responses to epinephrine, glucagon and insulin in the carp.' *Comptes Rendus Des Seances, Societé de Biologie et de Ses Filiales, 169* : 228-32.

Myers, S.F. and Avila, V.L. (1980) 'Tritiated 17β-oestradiol uptake by the brain and other tissues of the cichlid jewel fish, *Hemichromis bimaculatus.' General and Comparative Endocrinology, 42* : 203-11.

McBride, J.R., Fagerlund, U.H., Smith, M. and Tomlinson, N. (1964) 'Olfactory perception in juvenile salmon, II : conditioned response of juvenile sockeye salmon *(Oncorhynchus nerka)* to lake water.' *Canadian Journal of Zoology, 42 (2)* : 245-8.

McDonald, D.G. and Wood, C.M. (1981) 'Branchial and renal acid and ion fluxes in the rainbow trout *Salmo gairdneri* at low environmental pH.' *Journal of Experimental Biology, 93* : 101-18.

McEwen, B.S., Davis, P.G., Parsons, B. and Pfaff, D.W. (1979) 'The brain as a target for steroid hormone action.' *Annual Review of Neuroscience, 2* : 65-112.

McFarland, D. (ed.) (1981) *The Oxford Companion to Animal Behaviour*, University Press, Oxford, U.K.

McIntyre, D.C. and Healy, L.M. (1979) 'Effects of telecephalon damage on intraspecies aggression and activity in rainbow trout juveniles.' *Behavioural & Neural Biology, 25* : 490-501.

Mackay, I. and Mann, K.H. (1969) 'Fecundity of two cyprinid fishes in the River

Thames, Reading, England.' *Journal Fisheries Research Board of Canada, 26* : 2795-805.

MacQuarrie, D.W., Markert, J.R. and Vanstone, W.E. (1978) 'Photoperiod induced off-season spawning of coho salmon *(O. kisutch)*.' *Annales de biologie animale biochimie biophysique, 18* : 1051-8.

McVicar, A.J. (1982) *Renal and circulatory responses of the river lamprey (Lampetra fluviatilis L.) to changes in environmental salinity*, Unpublished Ph.D. Thesis, University of Wales, U.K.

Nakano, T. and Tomlinson, N. (1967) 'Catecholamine and carbohydrate concentrations in rainbow trout *(Salmo gairdneri)* in relation to physical disturbance.' *Journal of the Fisheries Research Board of Canada, 24* : 1701-15.

Nandi, J. (1964) 'The structure of the interrenal gland in teleost fishes.' *University of California Publications in Zoology, 65* : 129-95.

Nash, J. (1931) 'The number and size of glomeruli in the kidneys of fishes, with observations on the morphology of the renal tubules of fishes.' *American Journal of Anatomy, 47* : 425-45.

Nicol, J.A.C. (1952) 'Autonomic nervous systems in lower chordates.' *Biological Reviews, 27* : 1-49.

Niida, A., Oka, H. and Iwata, K.S. (1980) 'Visual responses of morphologically identified tectal neurons in the Crucian Carp.' *Brain Research, 201* : 361-71.

Nilsson, S. (1970) 'Excitatory and inhibitory innervation of the urinary bladder and gonads of a teleost *(Gadus morhua)*.' *Comparative and General Pharmacology, 1* : 23-28.

Nilsson, S. (1973) 'Fluorescent histochemistry of the catecholamines in the urinary bladder of a teleost *(Gadus morhua)*.' *Comparative and General Pharmacology, 4* : 17-21.

Nilsson, S. (1976) 'Fluorescent histochemistry a d cholinesterase staining of sympathetic ganglia in a teleost, *Gadus morhua.*' *Acta Zoologica 57* : 69-77.

Nilsson, S. (1983) *Autonomic nerve function in the vertebrates*. Springer Verlag, Berlin.

Nilsson, S. and Fänge, R. (1969) 'Adrenergic and cholinergic vagal effects on the stomach of a teleost *(Gadus morhua)*.' *Comparative Biochemistry and Physiology, 30* : 691-4.

Nilsson, S. and Grove, D.J. (1974) 'Adrenergic and cholinergic innervation of the spleen of the cod, *Gadus morhua*.' *European Journal of Pharmacology, 28* : 135-43.

Nilsson, S., Holmgren, S. and Grove, D.J. (1975) 'Effects of drugs and nerve stimulation on the spleen and arteries of two species of dogfish, *Scyliorhinus canicula* and *Squalus acanthias*.' *Acta Physiologica Scandinavica, 95* : 219-30.

Nilsson, S., Abrahamsson, T. and Grove, D.J. (1976) 'Sympathetic nervous control of adrenaline release from the head kidney of the cod, *Gadus morhua.*' *Comparative Biochemistry and Physiology, 55C* : 123-7.

Nilsson, S. and Pettersson, K. (1981) 'Sympathetic nervous control of blood flow in the gills of the Atlantic cod, *Gadus morhua*.' *Journal of Comparative Physiology, 144* : 157-63.

Nishimura, H. (1978) 'Physiological evolution of the renin-angiotensin system.' *Japanese Heart Journal, 19* : 8-822.

Nishimura, H and Sawyer, W.H. (1976) 'Vasopressor, diuretic and natriuretic responses to angiotensins by the American eel, *Anguilla rostrata*.' *General and Comparative Endocrinology, 29* : 337-48.

Nishimura, H., Norton, V.M. and Bumpus, F.M. (1978) 'Lack of specific inhibition of angiotensin II in eels by angiotensin antagonists.' *American Journal of Physiology, 235* : H95-103.

Nishimura, H. and Imai, M. (1982) 'Control of renal function in freshwater and

marine teleosts.' *Federation Proceedings, 41* : 2355-60.
Noaillac-Depeyre, J. and Hollande, E. (1981) 'Evidence for somatostatin, gastrin and pancreatic polypeptide-like substances in the mucosa cells of the gut in fishes with and without stomach'. *Cell Tissue Research, 216*: 193-203.
Noble, G.K., Kumpf, K.F. and Billings, V.N. (1938) 'The induction of brooding behaviour in the jewel fish.' *Journal of Endocrinology, 23*: 353-9.
Nomura, M. (1962) 'Studies on reproduction of rainbow trout *Salmo gairdneri*, with special reference to egg taking. 3. Acceleration of spawning by control of light.' *Bulletin of the Japanese Society of Scientific Fisheries, 28*: 1070-6.
Nordeng, H. (1971) 'Is the local orientation of anadromous fishes determined by pheromones?' *Nature 233*: 411-13.
Norris, H.W. (1941) *The plagiostome hypophysis, general morphology and types of structure,* Grinnell, Iowa, U.S.A.
Northcutt, R.G. and Bradford, M.R. (1978) 'New observations on the organisation and evolution of the telencephalon of actinopterygians' in S.O.E. Ebbesson (ed.), *Comparative Neurology of the Telencephalon,* Plenum Press, New York, pp. 41-98.
Northmore, D., Volkmann, F.C. and Yager, D. (1978) 'Vision in fishes: colour and pattern' in D.I. Mostofsky (ed.), *The Behaviour of Fish and Other Aquatic Animals,* Academic Press, New York.
O'Benar, J.D. (1976) 'Electrophysiology of neural units in the goldfish optic tectum.' *Brain Research Bulletin, 1* : 529-41.
Oide, H. and Utida, S. (1968) 'Changes in intestinal absorption and renal excretion of water during adaptation to sea-water in the Japanese eel.' *Marine Biology, 1* : 172-7.
Olivereau, M. (1954) 'Hypophyse et glande thyroide chez les poissons. Étude histophysiologique de quelques corrélations endocriniennes, en particulier chez *Salmo salar L.*' *Annales Institute Océanographique de Monaco, 29* : 95-296.
Olivereau, M. (1978) 'Les cellules gonadotropes chez les Salmonidés.' *Annales de biologie animale biochimie biophysique, 18* : 793-8.
Olivereau, M. (1980) 'Kidney structure and prolactin secretion in seawater eels treated with pimozide' in B. Lahlou (ed.), *Epithelial Transport in the Lower Vertebrates,* University Press, Cambridge, pp. 81-90.
Opdyke, D.F., Holcombe, R.F. and Wilde, D.W. (1979) 'Blood flow resistance in *Squalus acanthias.*' *Comparative Physiology and Biochemistry, 62A* : 711-17.
Opdyke, D.F., Carrol, R.G. and Keller, E. (1982) 'Catecholamine release and blood pressure changes induced by exercise in dogfish.' *American Journal of Physiology, 242*: R306-10.
Ormond, R.W. (1974) Visually responsive cells in the goldfish optic tectum, Unpublished Ph.D. Thesis, Cambridge University, U.K.
Oshima, K. and Gorbman, A. (1966a) 'Olfactory responses in the forebrain of goldfish and their modification by thyroxine treatment.' *General and Comparative Endocrinology, 7*: 398-409.
Oshima, K. and Gorbman, A. (1966b) 'Influence of thyroxine and sex steroids on the spontaneous and evoked unitary activity of the olfactory bulb of goldfish.' *General and Comparative Endocrinology, 7*: 482-91.
Oshima, K. and Gorbman, A. (1968) 'Modification by sex hormones of the spontaneous and evoked bulbar electrical activity in the goldfish.' *Journal of Endocrinology, 40*: 409-20.
Oshima, K. and Gorbman, A. (1969) 'Effect of oestradiol on NaCl-evoked olfactory bulbar potentials in goldfish : Dose-response relations.' *General and Comparative Endocrinology, 13*: 92-7.
Otsuka, N., Chihara, J., Sakurada, H. and Kanda, S. (1977) 'Catecholamine storing cells in the cyclostome heart.' *Archivum Histologicum Japonicum, 40* :

241-4.

Oyama, H., Martin, E.J., Sussman, K., Weir, G.C. and Permutt, A. (1981) 'The biological activity of catfish pancreatic somatostatin.' *Regulatory Peptides, 1* : 387-96.

Page, G.W. and Andrews, J.W. (1973) 'Interactions of dietary levels of protein and energy on channel catfish *(Ictalurus punctatus).*' *Journal of Nutrition, 103* : 1339-46.

Pang, P.K.T. and Sawyer, W.H. (1975) 'Parathyroid hormone preparations, salmon calcitonin, and urine flow in the South American lungfish, *Lepidosiren paradoxa.*' *Journal of Experimental Zoology, 193* : 407-12.

Palmer, T.N. and Ryman, B.E. (1972) 'Studies on oral glucose intolerance in fish.' *Journal of Fish Biology, 4* : 311-19.

Parker, G.H. (1948) *Animal colour changes and their neurohumours,* University Press, Cambridge, U.K.

Pärt, P., Tuurala, H. and Soivio, A. (1982) 'Oxygen transfer, gill resistance and structural changes in rainbow trout *(Salmo gairdneri)* (Richardson) gills perfused with vasoactive agents.' *Comparative Biochemistry and Physiology, 71C* : 7-13.

Partridge, B.L., Liley, N.R. and Stacey, N.E. (1976) 'The role of pheromones in the sexual behaviour of the goldfish.' *Animal Behaviour, 24* : 291-9.

Patent, G.J. and Foà, P.P. (1971) 'Radioimmunoassay of insulin in fishes, experiments *in vivo* and *in vitro.*' *General and Comparative Endocrinology, 16* : 41-6.

Peck, J.W. (1970) 'Straying and reproduction of coho salmon, *Oncorhynchus kisutch,* planted in a Lake Superior tributary.' *Transactions of the American Fisheries Society, 99* : 591-5.

Peter, R.E. (1979) 'The brain and feeding behaviour' in W.S. Hoar, D.J. Randall and J.R. Brett (eds.), *Fish Physiology, vol. VIII.* Academic Press, New York, pp. 121-59.

Peter, R.E., Monckton, E.A. and McKeown, B. (1976) 'The effects of gold thioglucose on food intake, growth and forebrain histology in goldfish, *Carassius auratus.*' *Physiology of Behaviour, 17* : 303-12.

Peter, R.E. and Hontela, A. (1978) 'Annual gonadal cycle in teleosts: environmental factors and gonadotrophin levels in blood.'in Assenmacher, I. and D.S. Farner (eds.), *Environmental Endocrinology,* Springer-Verlag, Berlin, pp. 20-25.

Pettersson, K. and Johansen, K. (1982) 'Hypoxic vasoconstriction and the effects of adrenaline on gas exchange efficiency in fish gills.' *The Journal of Experimental Biology, 97* : 263-72.

Peute, J., Goos, H.J.Th., Bruyn, M.G.A. and Oordt, P.G.W.J. van (1978) 'Gonadotropic cells of the rainbow trout pituitary during the annual cycle. Ultrastructure and hormone content.' *Annales de biologie animale biochimie biophysique, 18* : 905-10.

Peyraud-Waitzenegger, M. (1979) 'Simultaneous modifications of ventilation and arterial oxygen tension by catecholamines in the eel *Anguilla anguilla* L.: participation of α and β effects.' *Journal of Comparative Physiology B, 129* : 343-54.

Pfaff, D.W. (1980) *Oestrogens and brain function ; neural analysis of a hormone-controlled mammalian reproductive behaviour,* Springer, Berlin.

Pickering, A.D. (1976) 'Stimulation of intestinal degeneration by oestradiol and testosterone implantation in the migrating river lamprey, *Lampetra fluviatilis.*' *General and Comparative Endocrinology, 30* : 340-6.

Pickford, G.E. and Atz, J.W. (1957) *The physiology of the pituitary gland of fishes,* New York Zoological Society, U.S.A.

Pickford, G.E. and Strecker, E.L. (1977) 'The spawning reflex response of the killifish, *Fundulus heteroclitus*: isotocin is relatively inactive in comparison with arginine vasotocin.' *General and Comparative Endocrinology, 32* : 132-7.

Pickford, G.E. Knight, W.R. and Knight, J.M. (1980) 'Where is the spawning reflex receptor for neurohypophyseal peptides in the killifish, *Fundulus heteroclitus*?' *Revue canadienne de biologie, 39* : 97-105.

Piiper, J. and Scheid, P. (1977) 'Comparative physiology of respiration: functional analysis of gas exchange organs in vertebrates' in J.G. Widdicome (ed.), *International Review of Physiology, Respiratory Physiology II, Vol. 14.* University Park Press, Baltimore, U.S.A., pp. 219-53.

Piiper, J., Meyer, M., Worth, H. and Willmer, H. (1977) 'Respiration and circulation during swimming activity in the dogfish *Scyliorhinus stellaris.*' *Respiration Physiology, 30* : 221-39.

Pitcher, T.J. (1979) 'Sensory information and the organisation of behaviour in a shoaling cyprinid fish.' *Animal Behaviour, 27* : 126-49.

Pitcher, T.J., Kennedy, G.J.A. and Wirjoatmodjo, S. (1979) 'Links between the ecology and behaviour of fishes.' *Proceedings of British Freshwater Fisheries Conference, 1* : 162-75.

Plisetskaya, E.M. (1968) 'Brain and heart glycogen content in some vertebrates, and effect of insulin.' *Endocrinologia Experimentia, 2* : 251-62.

Plisetskaya, E.M. (1980) 'Fatty acid levels in blood of cyclostomes and fish.' *Environmental Biology and Fisheries, 5* : 273-90.

Plisetskaya, E.M., Leibush, B.N. and Bondareva, V. (1976) 'The secretion of insulin and its role in cyclostomes and fishes' in T.A.I. Grillo, L. Leibson & A. Epple (eds.), *The Evolution of Pancreatic Islets,* Pergamon Press, Oxford, U.K. pp. 251-69.

Plisetskaya, E.M., Soltitskaya, L.P. and Leibson, L.G. (1977) 'Insulin in the blood of diadromous lampreys and fish on various stages of spawning migration' in E. Plisetskaya and L. Leibson (eds.), *Evolutionary Endocrinology of the Pancreas,* Nauka, Leningrad, pp. 127-133 (In Russian).

Porter, K.R. and Tucker, J.B. (1981) 'The ground substance of the living cell.' *Scientific American, 244* : 40-51.

Potson, H.A. and Livingston, D.L. (1971) 'The effect of continuous darkness and continuous light on the functional sexual maturity of brook trout during their second reproductive cycle.' *Fisheries Research Bulletin, New York 33* : 25-29 (Cortland Hatchery Report No. 38, for the year 1969).

Purdom, C.E. (1972) 'Induced polyploidy in plaice *(Pleuronectes platessa)* and its hybrid with the flounder *(Platichthys flesus).*' *Heredity, 29* : 11-23.

Purdom, C.E. (1979) 'Genetics of growth and reproduction in teleosts.' *Symposium of the Zoological Society of London, 44* :207-17.

Putten, L.J.A. van, Peute, J., Oordt, P.G.W.J. van, Goos, J.H.Th. and Breton, B. (1981) 'Glycoprotein gonadotrophin in the plasma and its cellular origin in the adenohypophysis of sham-operated and ovariectomised rainbow trout, *Salmo gairdneri.*' *Cell and Tissue Research, 218* : 439-48.

Pyle, E.A. (1969) 'The effect of constant light or constant darkness on the growth and sexual maturity of brook trout.' *Fisheries Research Bulletin, New York, 31* : 13-19 (Cortland Hatchery Report No. 36, for the year 1967).

Randall, D. (1982) 'The control of respiration and circulation in fish during exercise and hypoxia'. *The Journal of Experimental Biology, 100* : 275-88.

Randall, D.J., Holeton, G.F. and Stevens, E.D. (1967) 'The exchange of oxygen and carbon dioxide across the gills of rainbow trout.' *The Journal of Experimental Biology, 46* : 339-48.

Rankin, J.C. and Davenport, J.D. (1981) *Animal Osmoregulation,* Blackies, Glasgow, U.K.

Rankin, J.C. and Babiker, M.M. (1981) 'Circulatory effects of catecholamines in the eel, *Anguilla anguilla L.*' in A.D. Pickering (ed.), *Stress and Fish,*, Academic Press, New York, pp. 352-3.

Rankin, J.C., Logan, A.G. and Moriarty, R.J. (1980) 'Changes in kidney function in the river lamprey, *Lampetra fluviatilis L.*, in response to changes in external salinity' in B. Lahlou, *Epithelial Transport in the Lower Vertebrates*, University Press, Cambridge, pp. 171-84.

Rankin, J.C., Griffiths, V., McVicar, A.J. and Gilham, I.D. (1983) 'Control of kidney function in the river lamprey.' *Symposium Proceedings of the Ninth International Symposium on Comparative Endocrinology, Hong Kong, 1981.* (In press).

Rasa, O.A.E. (1969) 'Territoriality and the establishment of dominance by means of visual cues in *Pomacentrus jenkinsii.*' *Zeitschrift für Tierpsychologie 26*: 825-45.

Rasa, O.A.E. (1971) 'The causal factors and function of 'yawning' in *Microspathodon chrysurus.* (Pisces : Pomacentridae).' *Behaviour 39*: 39-57.

Rasa, O.A.E. (1981). Entry on 'Aggression' in Mcfarland (1981) q.v.

Read, J.B. and Burnstock, G. (1968a) 'Comparative histochemical studies of adrenergic nerves in the enteric plexuses of vertebrate large intestine.' *Comparative Biochemistry & Physiology, 27*: 505-17.

Read, J.B. and Burnstock, G. (1968b) 'Fluorescent histochemical studies on the mucosa in the vertebrate gastrointestinal tract.' *Histochemie, 16*: 324-32.

Read, J.B. and Burnstock, G. (1969) 'Adrenergic innervation of the gut musculature in vertebrates.' *Histochemie, 17*: 263-92.

Reinboth, R. (1972) 'Some remarks on secondary sex characters, sex and sexual behaviour in teleosts.' *General & Comparative Endocrinology Supplement, 3*: 565-70.

Reinecke, M., Carraway, R.E., Falkner, S., Feurle, G.E. and Forssmann, W.G. (1980) 'Occurrence of neurotensin-immunoreactive cells in the digestive tract of lower vertebrates and deuterostomian invertebrates. A correlated immunohistochemical and radioimmunochemical study.' *Cell Tissue Research, 212*: 173-84.

Reite, O.B. (1969) 'The evolution of vascular smooth muscle responses to histamine and 5-hydroxytryptamine. I. Occurrence of stimulatory actions in fish.' *Acta Physiologica Scandinavica, 75*: 221-39.

Riegel, J. (1978) 'Factors affecting glomerular functions in the Pacific hagfish. *Eptatretus stouti* (Lockington). *Journal of Experimental Biology, 73*: 261-77.

Riehl, R (1978) 'Elektronenmikroskopische und autoradiographische Untersuchungen an den Dotterkernen in den Oocyten von *Noemacheilus barbatulus (L.)* und *Phoxinus phoxinus (L.)* (Pisces, Teleostei).' *Cytobiologie, 17*: 137-45.

Robertson, O.H. and Wexler, B.C. (1960) 'Histological changes in the organs and tissues of migrating and spawning Pacific salmon (genus *Oncorhynchus).*' *Endocrinology, 66*: 222-39.

Rodieck, R.W. (1973) *The Vertebrate Retina*, Freeman, San Francisco.

Rombout, J.H.W.M. (1977) 'Enteroendocrine cells in the digestive tract of *Barbus conchonius* (Teleostei, Cyprinidae).' *Cell Tissue Research, 185*: 435-50.

Ronner, P. and Scarpa, A. (1981) 'Release of glucagon and somatostatin from the perfused Brockmann body of channel catfish.' *Federation Proceedings 30,* 1617 (Abstract No. 451).

Rouse, E.F., Coppenger, C.J. and Barnes, P.R. (1977) 'The effect of an androgen inhibitor on behaviour and testicular morphology in the stickleback, *Gasterosteus aculeatus.*' *Hormones & Behaviour, 9*: 8-18.

Rounsefell, G.A. and Kelez, G.B. (1938) 'The salmon and salmon fisheries of

Swiftsure Bank, Puget Sound and the Fraser River.' *Bulletin U.S. Bureau of Fisheries, 49*: 693-823.

Rowell, T.E. (1974) 'The concept of social dominance.' *Behavioural Biology, 11*: 131-54.

Rushmer, R.F. (1970) *Cardiovascular Dynamics*. W.B. Saunders Co., London.

Saito, K. (1973) 'Nervous control of intestinal motility in goldfish.' *Japanese Journal of Smooth Muscle Research, 9*: 79-86.

Sanchez-Rodriguez, M., Escaffre, A.M., Marlot, S. and Reinaud, P. (1978) 'The spermiation period in the rainbow trout (*Salmo gairdneri*). Plasma gonadotrophin and androgen levels, sperm production and biochemical changes in the seminal fluid.' *Annales de biologie animale biochimie biophysique, 18*: 943-8.

Sandeman, D.C. and Rosenthal, N.P. (1974) 'Efferent axons in the fish optic nerve and their effect on the retinal ganglion cells.' *Brain Research, 68*: 41-54.

Sangalang, G.B., Freeman, H.C. and Flemming, R.B. (1978) 'A simple technique for determining the sex of fish by radio-immunoassay using 11-ketotestosterone antiserum.' *General and Comparative Endocrinology, 36*:187-93.

Santer, R.M. (1977) 'Monoaminergic nerves in the central and peripheral nervous systems of fishes.' *General and Comparative Pharmacology, 8*: 157-72.

Satake, N. (1979) 'Melatonin mediation in sedative effect of serotonin in goldfish.' *Physiology of Behaviour, 22*: 817-19.

Satchell, G.H. (1961) 'The response of the dogfish to anoxia.' *Journal of Experimental Biology, 38*: 531-43.

Satchell, G.H. (1968) 'A neurological basis for the co-ordination of swimming with respiration in fish.' *Comparative Biochemistry and Physiology, 27*: 835-41.

Satchell, G.H. (1971) *Circulation in Fishes*, University Press, Cambridge, U.K.

Savage, G.E. and Roberts, M.G. (1975) 'Behavioural effects of electrical stimulation of the hypothalamus of the goldfish.' *Brain, Behaviour and Evolution, 12*: 42-56.

Sawyer, W.H. (1970) 'Vasopressor, diuretic and natriuretic responses by lungfish to arginine vasotocin.' *American Journal of Physiology, 218*:1789-94.

Sawyer, W.H., Uchiyama, M. and Pang, P.K.T. (1982) 'Control of renal function in lungfishes.' *Federation Proceedings, 41*: 2361-4.

Sawyer, W.H., Blair-West, J.R., Simpson, P.A. and Sawyer, M.K. (1976) 'Renal responses of Australian lungfish to vasotocin, angiotensin II and NaCl infusion.' *American Journal of Physiology, 231*: 593-602.

Scanes, C.G., Dobson, S., Follett, B.K. and Dodd, J.M. (1972) 'Gonadotrophic activity in the pituitary gland of the dogfish (*Scyliorhinus canicula*).' *Journal of Endocrinology, 54*: 343-4.

Schaeffer, B. and Williams, M. (1977) 'Relationship of fossil and living elasmobranchs.' *American Zoologist, 17*: 213-302.

Scheer, B.T. and Langford, R.W. (1976) 'Endocrine effects on the cation-dependent ATPase of the gills of European eels (*Anguilla anguilla L.*), and efflux of sodium'. *General and Comparative Endocrinology, 30*: 313-26.

Schellart, N.A.M. and Spekreijse, H. (1976) 'Shapes of receptive field centers in optic tectum of goldfish.' *Vision Research, 16*: 1018-20.

Schmidt, J.T. (1979) 'The laminar organization of optic nerve fibres in the tectum of the goldfish.' *Proceedings of the Royal Society, London, B. 205*: 287-306.

Schmidt, P.J. and Idler, D.R. (1962) 'Steroid hormones in the plasma of salmon at various stages of maturation.' *General and Comparative Endocrinology, 2*: 204-14.

Schmidt-Nielsen, B. and Renfro, J.L. (1975) 'Kidney function of the American eel *Anguilla rostra.' American Journal of Physiology 228*: 420-31.

Schmidt-Nielsen, B., Truniger, B. and Rabinowitz. L. (1972) 'Sodium-linked urea transport by renal tubules of spiny dogfish, *Squalus acanthias.' Comparative Biochemistry and Physiology 42A* : 13-25.

Scholz, A.T., Cooper, J.C., Madison, D.M. Horrall, R.M. Hasler, A.D., Dizon, A.E. and Poff, R.J. (1973) 'Olfactory imprinting in coho salmon : behavioral and electrophysiological evidence.' *Proceedings of the Conference on Great Lakes Research, 16* : 143-53.

Scholz, A.T., Horrall, R.M. Cooper, J.C. and Hasler, A.D. (1975) 'Imprinting to chemical cues; the basis for homestream selection in salmon.' *Science, 196* : 1247-9.

Scholz, A.T., Horrall, R.M., Cooper, J.C. and Hasler, A.D. (1978a) 'Homing of morpholine imprinted brown trout *(Salmo trutta).' Fishery Bulletin (Seattle), 76(1)* : 293-5.

Scholz, A.T., Gosse, C.K., Cooper, J.C., Horrall, R.M. Hasler, A.D. Daly, R.I. and Poff, R.J. (1978b) 'Homing of rainbow trout transplanted in Lake Michigan: a comparison of three procedures used for imprinting and stocking.' *Transactions of the American Fisheries Society, 107(3)* : 439-43.

Schreck, C.B. (1972) 'Evaluation of diel variation in androgen levels of rainbow trout, *Salmo gairdneri.' Copeia 1972(4)* : 865-8.

Schreck, C.B. (1981) 'Stress and compensation in teleostean fishes : response to social and physical factors' in A.D. Pickering (ed.), *Stress and Fish,* Academic Press, New York, pp. 295-321.

Schreck, C.B., Lackey, R.T. and Hopwood, M.L. (1973) 'Plasma oestrogen levels in rainbow trout, *Salmo gairdneri.' Journal of Fish Biology, 5* : 227-30.

Scott, A.P. (1981) 'Imports — time for British farms to plug the widening gap.' *Fish Farmer, 4* : 28-9.

Scott, A.P. and Baynes, S.M. (1980) 'A review of the biology, handling and storage of salmonid spermatozoa.' *Journal of Fish Biology, 17* : 707-39.

Scott, A.P., Bye, V.J. and Baynes, S.M. (1980) 'Seasonal variations in sex steroids of female rainbow trout *(Salmo gairdneri).' Journal of Fish Biology, 17* : 587-92.

Scott, A.P., Sheldrick, E.L. and Flint, A.P.F. (1982) 'Measurement of 17,20-dihyrodxy-4-pregnen-3-one in plasma of trout *(Salmo gairdneri).* Seasonal changes and response to salmon pituitary extract.' *General and Comparative Endocrinology, 46* : 444-51.

Scott, A.P., Sumpter, J.P. and Hardiman, P.A. (1983) 'Hormone changes during ovulation in the rainbow trout *(Salmo gairdneri).' General and Comparative Endocrinology, 49* : 128-134.

Scott, A.P., Bye, V.J., Baynes, S.M. and Springate, J.R.C. (1980) 'Seasonal variations in plasma concentrations of 11-ketotestosterone and testosterone in male rainbow trout *(Salmo gairdneri).' Journal of Fish Biology, 17* : 495-505.

Scott, J.P. (1981) 'The evolution of function in agonistic behaviour' in P.F. Brain and D. Benton (eds.), *Multidisciplinary approaches in aggression research,* Elsevier/North Holland, Amsterdam, pp. 129-56.

Segaar, J. (1961) 'Telencephalon and behaviour in *Gasterosteus aculeatus.' Behaviour, 18* : 256-87.

Sevenster, P. and Goyens, J. (1975) 'Experience sociale et rôle du sexe dans l'ontogènese de l'aggressivité chez l'épinoche *(Gasterosteus aculeatus L.).' Netherlands Journal of Zoology, 25* : 195-205.

Shackley, S.E. and King, P.E. (1977) 'The reproductive cycle and its control; frequency of spawning and fecundity in *Blennius pholis L.' Journal of Experimental Marine Biology and Ecology, 30* : 73-83.

Shapiro, D.Y. and Boulon, R.H. (1982) 'The influence of females on the initiation of female-to-male sex change in a coral reef fish.' *Hormones & Behaviour, 16* : 66-75.

Shelton, G. (1970) 'The regulation of breathing' in W.S. Hoar and D.J. Randall (eds.) *Fish Physiology, Vol. IV*, Academic Press, New York, pp. 293-359.

Shiraishi, Y and Fukuda, Y. (1966) 'The relation between the daylength and the maturation in four species of salmonid fish.' *Bulletin of the Freshwater Fisheries Research Laboratory of Tokyo, 16*: 103-111.

Short, S., Butler, P.J. and Taylor, E.W. (1977) 'The relative importance of nervous, humoral and intrinsic mechanisms in the regulation of heart rate and stroke volume in the dogfish *Scyliorhinus canicula.' The Journal of Experimental Biology, 70*: 77-92.

Simpson, T.H., Wright, R.S. and Hunt, S.V. (1963) 'Sex hormones in fish. II. The oestrogens of *Scyliorhinus caniculus.' Journal of Endocrinology, 26*: 499-507.

Simpson, T.H., Wright, R.S. and Hunt, S.V. (1964) 'Steroid biosynthesis in the testis of dogfish *(Squalus acanthias).' Journal of Endocrinology, 31*: 29-38.

Simpson, T.H. and Wardle, C.S. (1967) 'A seasonal cycle in the testis of the spurdog, *Squalus acanthias*, and the site of 3ß-hydroxy-steroid dehydrogenase activity.' *Journal of the Marine Biological Association of U.K. 47*: 699-708.

Singley, J.A. and Chavin, W. (1976) 'Alterations in goldfish hormone levels during reproductive activity.' *American Zoologist, 16*: 258 (Abstract).

Sire, O. and Dépêche, J. (1981) '*in vitro* effect of a fish gonadotrophin on aromatase and 17β-hydroxysteroid dehydrogenase activities in the ovary of the rainbow trout *(Salmo gairdneri). Reproduction Nutrition Développment, 21*: 715-26.

Sivak, J.G. (1973) 'Accommodation in some species of North American fishes.' *Journal of the Fisheries Research Board of Canada, 30*: 1141-6.

Slater, P.J.B. (1978) *Sex hormones and behaviour*, E. Arnold, London.

Smith, C.L. (1975) 'The evolution of hermaphroditism in fish' in R. Reinboth (ed.) *Intersexuality in the Animal Kingdom*, Springer-Verlag, Berlin, pp. 295-310.

Smith, D.G. (1978) 'Neural regulation of blood pressure in rainbow trout *(Salmo gairdneri).' Canadian Journal of Zoology, 56*: 16/8-83.

Smith, P.H. and Porte, D., Jr. (1976) 'Neuropharmacology of the pancreatic islets.' *Annual Review of Pharmacology and Toxicology, 16*: 269-85.

Smith, R.F.J. (1969) 'Control of pre-spawning behaviour of sunfish *(Lepomis gibbosus* and *L. megalotis)*. I. Gonadal androgen.' *Animal Behaviour, 17*: 279-85.

Smith, R.J.F. (1970) 'Control of prespawning behaviour of sunfish (*Lepomis gibbosus* and *L.megalotis)*. II. Environmental factors.'*Animal Behaviour, 18*: 575-87.

Smith, R.J.F. and Hoar, W.S. (1967) 'The effects of prolactin and testosterone on the parental behaviour of the male stickleback, *Gasterosteus aculeatus.' Animal Behaviour, 15*: 342-52.

Sönksen, P.H., Jones, R.H., Tompkins, C.V., Srivastava, M.C. and Nabarro, J.D.N. (1976) 'The Metabolism of insulin *in vivo*'. *Excerpta Medica International Congress (Series 413)*, pp 204-212 Elsevier, Amsterdam.

Spiliotis, P.H. (1974) 'The effect of thyroxine and thiourea on territorial behaviour in cichlid fish'. *Dissertation Abstracts International, 35B*: 1030.

Stabrovski, E.M. (1969) 'Adrenaline and noradrenaline in the organs of elasmobranch (cartilaginous) and teleost (bony) fish of the Black Sea.' *Journal of Evolutionary Biochemistry and Physiology, 5*: 38-41.

Stacey, N.E. (1976) 'Effects of indomethacin and prostaglandins on spawning behaviour of female goldfish.' *Prostaglandins, 12*: 113-26.

Stacey, N.E. (1977) *The Regulation of Spawning Behaviour in the Female Goldfish, Carassius auratus*, Unpublished Ph.D. Thesis, University of British Columbia.

Stacey, N.E. (1981) 'Hormonal regulation of female sexual behaviour in teleosts.'

282　　*References*

American Zoologist, 21 : 305-16.

Stacey, N.E. and Liley, N.R. (1974) 'Regulation of spawning behaviour in the female goldfish.' *Nature, 247* : 71-2.

Stacey, N.E. and Peter, R.E. (1979) 'Central action of prostaglandins in spawning behaviour of female goldfish.' *Physiology and Behaviour, 22* : 1191-6.

Stacey, N.E., Cook, A.F. and Peter, R.E. (1979) 'Ovulatory surge of gonadotropin in the goldfish.' *General and Comparative Endocrinology, 37* : 246-9.

Stacey, N.E. and Goetz, F.W. (1982) 'Role of prostaglandins in fish reproduction.' *Canadian Journal of Fisheries and Aquatic Sciences, 39* : 92-8.

Stacey, P.B. and Chiszar, D. (1978) 'Body colour pattern and aggressive behaviour of male pumpkinseed sunfish *(Lepomis gibbosus)*'. *Behaviour, 64* : 271-304.

Stanley, H.P. (1966) 'The structure and development of the seminiferous follicle in *Sycliorhinus caniculus* and *Torpedo marmorata* (Elasmobranchii)'. *Zeitschrift für Zellforschung, 75* : 453-68.

Stanley, H.P. (1971a) 'Fine structure of spermiogenesis in the elasmobranch fish *Squalus suckleyi*. I. Acrosome formation, nuclear elongation and differentiation of the midpiece axis.' *Journal of Ultrastructure Research, 36* : 86-102.

Stanley, H.P. (1971b) 'Fine structure of spermiogenesis in the elasmobranch fish *Squalus suckleyi*. II. Late stages of differentiation and structure of the mature spermatozoon.' *Journal of Ultrastructure Research, 36* : 103-18.

Stannius, H. (1849) *Das Peripherische Nervensystem der Fische*, Stiller, Rostock.

Stasko, A.B. (1971) 'Review of field studies on fish orientation.' *Annals of the New York Academy of Sciences, 188* : 12-29.

Steen, J.B. and Kruysse, A. (1964) 'The respiratory function of teleostean gills.' *Comparative Biochemistry and Physiology, 12* : 127-42.

Stein, D.L. (1980) 'Aspects of reproduction of liparid fishes from the continental slope and abyssal plain off Oregon, with notes on growth.' *Copeia, 1980* : 687-99.

Stene-Larsen, G. (1981) 'Comparative aspects of cardiac adrenoceptors: characterization of the β_2-adrenoceptor as a common "adrenaline" receptor in vertebrate hearts.' *Comparative Biochemistry and Physiology, 70C* : 1-12.

Stevenson, S.V. and Grove, D.J. (1977) 'The extrinsic innervation of the stomach of the plaice, *Pleuronectes platessa* L. I — the vagal nerve supply.' *Comparative Biochemistry and Physiology, 58C* : 143-51.

Stevenson, S.V. and Grove, D.J. (1978) 'The extrinsic innervation of the stomach of the plaice, *Pleuronectes platessa* L. II — The splanchnic nerve supply.' *Comparative Biochemistry and Physiology, 60C* : 45-50.

Stuart, T.A. (1959) 'Tenth annual report of the supervisory committee for brown trout research, 1957-1958.' *Freshwater and Salmon Fisheries Research, 23* : 6-7.

Sumpter, J.P. (1983) 'The purification, radioimmunoassay and plasma levels of vitellogenin from the rainbow trout *Salmo gairdneri*.' *Proceedings of the Ninth International Symposium on Comparative Endocrinology, Hong Kong, 7th-11th December, 1981.* (In press).

Sumpter, J.P. and Dodd, J.M. (1979) 'The annual reproductive cycle of the female lesser spotted dogfish, *Scyliorhinus canicula* L. and its endocrine control.' *Journal of Fish Biology, 15* : 687-95.

Sumpter, J.P., Follett, B.K. and Dodd, J.M. (1978a) 'Studies on the purification of gonadotrophin from dogfish *(Scyliorhinus canicula, L.)* pituitary glands.' *Annales Biologique Animale Biochimique et Biophysique. 18(4)* : 787-91.

Sumpter, J.P., Follett, B.K., Jenkins, N. and Dodd, J.M. (1978b) 'Studies on the purification and properties of gonadotrophin from ventral lobes of the pituitary gland of the dogfish *(Scyliorhinus canicula L.)*.' *General and Comparative Endocrinology, 36* : 264-74.

Sundararaj, B.I. (1979) 'Some aspects of oocyte maturation in catfish.' *Journal of Steroid Biochemistry, 11* : 701-7.

Sundararaj, B.I. (1981) *Reproductive physiology of teleost fishes,* Food and Agriculture Organization of the United Nations, ADCP/REP/81/16.

Sutterlin, A.M. and Prosser, C.L. (1970) 'Electrical properties of the goldfish optic tectum.' *Journal of Neurophysiology, 33* : 36-45.

Taborsky, M. and Limberger, D. (1981) 'Helpers in fish.' *Behavioural Ecology & Sociobiology, 8* : 143-5.

Tarrant, R.M. (1966) 'Thresholds of perception of eugenol in juvenile salmon.' *Transactions of the American Fisheries Society, 95* : 112-15.

Tashima, L. and Cahill, G.F., Jr. (1968) 'Effects of insulin in toadfish *Opsanus tau.' General and Comparative Endocrinology, 11* : 262-71.

Tavolga, W.N. (1955) 'Effects of gonadectomy and hypophysectomy on pre-spawning behaviour in males of the gobiid fish *Bathygobius soporator.' Physiological Zoology, 28* : 218-33.

Taylor, E.W. and Butler, P.J. (1982) 'Nervous control of heart rate and activity in the cardiac vagus of the dogfish.' *Journal of Applied Physiology, 53* : 1330-35.

Taylor, E.W., Short, S. and Butler, P.J. (1977) 'The role of the cardiac vagus in the response of the dogfish *Scyliorhinus canicula* to hypoxia.' *The Journal of Experimental Biology, 70* : 57-75.

Templeman, W. (1944) 'The life history of the spiny dogfish and the vitamin A values of the dogfish liver oil.' *Dep. Nat. Res. Nfld. Res. Bull. 15* : 1-102.

Teshima, K. (1981) 'Studies on the reproduction of Japanese smooth dogfishes, *Mustelus manazo* and *M. griseus.' Journal of Shimonoseki University of Fisheries, 29* : 113-99.

Teshima, K. and Mizue, K. (1972) 'Studies on sharks. I. Reproduction in the female sumitsuki shark, *Carcharhinus dussumieri.' Marine Biology, 14* : 222-31.

Te Winkel, L.E. (1943) 'Observations on later phases of embryonic nutrition in *Squalus acanthias.' Journal of Morphology, 73* : 177-205.

Te Winkel, L.E. (1969) 'Specialised hasophilic cells in the ventral lobe of the pituitary of the smooth dogfish *Mustelus canis.' Journal of Morphology, 127* : 439-52.

Thorgaard, G.H. and Gall, G.A.E. (1979) 'Adult triploids in a rainbow trout family.' *Genetics, 93* : 961-73.

Thorpe, A. (1976) 'Studies on the role of insulin in teleost metabolism' in T.A.I. Grillo, L. Leibson and A. Epple (eds.), *The Evolution of Pancreatic Islets,* Pergamon Press, Oxford, U.K., pp. 271-284.

Thorpe, A. and Ince, B.W. (1976) 'Plasma insulin levels in teleosts determined by a charcoal-separation radioimmunoassay technique.' *General and Comparative Endocrinology, 30* : 332-9.

Thorpe, A. and Duve, H. (1980) 'Isolated islets of Langerhans of the silver eel *(Anguilla anguilla)* in culture.' *General and Comparative Endocrinology, 40* : 351-2.

Tokartz, R.R. (1978) 'Oogonial proliferation, oogenesis, and folliculogenesis in non-mammalian vertebrates' in R.E. Jones (ed.) *The Vertebrate Ovary,* Plenum Press, New York, pp. 145-79.

Tokartz, R.R. and Crews, D. (1981) 'Effects of prostaglandins on sexual receptivity in the female lizard, *Anolis carolinensis.' Endocrinology, 109* : 451-7.

Toury, R.J., Clérot, C. and André, J. (1977) 'Les groupements mitochondriaux des cellules germinals des poissons Téléostéen Cyprinidés. IV. Analyse biochimique des constituants du "ciment" intermitochondrial isolé.' *Biologie Cellulaire, 30* : 225-31.

Tuge, H. and Uchihashi, N. (1968) *An atlas of the brains of fishes of Japan,* Tsujiki Shokan, Tokyo.

Tyler, A.V. and Dunn, R.S. (1976) 'Ration, growth and measures of somatic and organ condition in relation to the meal frequency in winter flounder, *Pseudopleuronectes americanus,* with hypotheses regarding population homeostasis.' *Journal of the Fisheries Research Board of Canada, 33* : 63-75.

Umminger, B.L., Benziger, D. and Levy, S. (1975) '*In vitro* stimulation of hepatic glycogen phosphorylase activity by epinephrine and glucagon in the killifish, *Fundulus heteroclitus.' Comparative Biochemistry and Physiology, 51* : 111-116.

Unger, R.H., Dobbs, R.E. and Orci, L. (1978) 'Insulin, glucagon and somatostatin secretion in the regulation of metabolism.' *Annual Review of Physiology, 40* : 307-43.

Urist, M.R. and Schjeide, O.A. (1961) 'The partition of calcium and protein in the blood of oviparous vertebrates during estrus.' *Journal of General Physiology, 44* : 743-56.

Vahl, O. (1979) 'An hypothesis on the control of food intake of fish.' *Aquaculture, 17* : 221-9.

VanderWeele, D.A. and Sanderson, D. (1976) 'Peripheral glucosensitive satiety in the rabbit and the rat' in D. Navin, W. Wyrivicka and G. Bray (eds.), *Hunger : Basic mechanisms and clinical implications,* Raven Press, New York, pp. 383-93.

Vanegas, H., Amat, J. and Essayag-Millan, E. (1974) 'Post-synaptic phenomena in optice tectum neurons following optic nerve stimulation in fish.' *Brain Research, 77* : 25-38.

Vanegas, H. (1975) 'Cytoarchitecture and connections of the teleostean optic tectum' in M. Ali (ed.), *Vision in Fishes,* Plenum, New York, pp. 151-8.

Vanegas, H., Williams, B. and Freeman, J.A. (1979) 'Responses to simulation of the marginal fibres in the teleostean optic tectum.' *Experimental Brain Research, 34* : 335-49.

Van Noorden, S. and Patent, G.J. (1980) 'Vasoactive intestinal polypeptide-like immunoreactivity in nerves of the pancreatic islet of the teleost fish, *Gillichthys mirabilis.' Cell Tissue Research, 212* : 139-46.

VanRee, G.E., Lok, D. and Bosman, G.J.C.G.M. (1977) '*In vitro* induction of nuclear breakdown in oocytes of the zebrafish, *Brachydanio rerio* (Ham. Buch.). Effects of composition of the medium and of protein and steroid hormones.' *Proceedings Nederlandse Akademie van Wetenschappen, 80C* : 353-71.

Villars, T.A. and Davis, R.E. (1977) 'Castration and reproductive behavior in the paradise fish, *Macropodus opercularis (L.)* (Osteochthyes : Belontiidae).' *Physiology and Behavior, 19* : 371-5.

Vlad, M. (1976) 'Nucleolar DNA in oocytes of *Salmo irideus* (Gibbons).' *Cell and Tissue Research, 167* : 407-24.

Vodegel, N. (1978) 'A study of the motivation underlying some communicative behaviours of *Pseudotropheus zebra* (Pisces, Cichlidae); a mathematical model.' *Proc. Kon. Ned. Akad. v. Wet., Ser.* C, *81* : 211-40.

Wablin, F.S. and Dowling, J.E. (1969) 'Organization of the retina of the mud-puppy *Necturus maculatus.* II Intracellular recording.' *Journal of Neurophysiology, 32* : 339-55.

Wahlqvist, I. (1980) 'Effects of catecholamines on isolated systemic and branchial vascular beds of the cod, *Gadus morhua.' Journal of Comparative Physiology, 137* : 139-43.

Wahlqvist, I. (1981) 'Branchial vascular effects of catecholamines released from the head kidney of the Atlantic cod, *Gadus morhua.' Molecular Physiology, 1* :

235-41.

Wahlqvist, I., and Nilsson, S. (1977) 'The role of sympathetic fibres and circulating catecholamines in controlling the blood pressure and heart rate in cod, *Gadus morhua*.' *Comparative Biochemistry and Physiology, 56C* : 65-7.

Wahlqvist, I. and Nilsson, S. (1980) 'Adrenergic control of the cardio-vascular system of the Atlantic cod, *Gadus morhua* during "stress".' *Journal of Comparative Physiology, 137* : 145-50.

Wahlqvist, I. and Nilsson, S. (1981) 'Sympathetic nervous control of the vasculature in the tail of the Atlantic cod, *Gadus morhua*.' *Journal of Comparative Physiology, 144* : 153-6.

Wai, E.H. and Hoar, W.S. (1963) 'The secondary sex characters and reproductive behaviour of gonadectomised sticklebacks treated with methyltestosterone.' *Canadian Journal of Zoology, 41* : 611-28.

Walker, J.C. (1967) *Odor discrimination in relation to homing in Atlantic salmon.* Unpublished M.Sc. thesis, University of New Brunswick, Fredericton, Canada.

Wallace, R.A. (1978) 'Oocyte growth in non-mammalian vertebrates' in R.E. Jones (ed.), *The Vertebrate Ovary*, Plenum Press, New York, pp. 469-502.

Wallace, R.A. and Selman, K. (1979) 'Physiological aspects of oogenesis in two species of sticklebacks, *Gasterosteus aculeatus L.* and *Apeltes quadracus* (Mitchill).' *Journal of Fish Biology, 14* : 551-64.

Wallace, R.A. and Selman, K. (1981) 'Cellular and dynamic aspects of oocyte growth in teleosts.' *American Zoologist, 21* : 325-43.

Walton, M.J. and Cowey, C.B. (1979) 'Gluconeogenesis by isolated hepatocytes from rainbow trout, *Salmo gairdneri*.' *Comparative Biochemistry and Physiology, 62B* : 75-9.

Wapler-Leong, D.C.Y. and Reinboth, R. (1974) 'The influence of androgenic hormones on the behaviour of *Haplochromis burtoni* (Cichlidae).' *Fortschrifte des Zoologie 22* : 334-9.

Wartzok, D. and Marks, W.B. (1973) 'Directionally selective visual units recorded in the optic tectum of the goldfish.' *Journal of Neurophysiology, 36* : 588-604.

Watkins, D.T. (1972) 'Pyridine nucleotide stimulation of insulin release from isolated toadfish insulin secretion granules.' *Endocrinology, 90* : 272-6.

Watson, A.H.D. (1979) 'Fluorescent histochemistry of the teleost gut; evidence for the presence of serotonergic neurones.' *Cell Tissue Research, 197* : 155-64.

Watson, A.H.D. (1981) 'The ultrastructure of the innervation of the intestinal wall in the teleosts *Myoxocephalus scorpius* and *Pleuronectres platessa*.' *Cell Tissue Research, 214* : 651-8.

Wedemeyer, G. (1972) 'Some physiological consequences of handling stress in the juvenile Coho salmon *(Oncorhyndrus kisutch)* and steelhead trout *(Salmo gairdneri).' Journal of the Fisheries Research Board of Canada, 29* : 1780-3.

Weis, C.S. and Coughlin, J.P. (1979) 'Maintained aggressive behaviour in gonadectomised male Siamese fighting fish *(Betta splendens)*.' *Physiology of Behaviour, 23* : 173-7.

Wendelaar-Bonga, S.E. (1976) 'The effect of prolactin on kidney structure of the euryhaline teleost *Gasterosteus aculeatus* during adaptation to freshwater.' *Cell and Tissue Research, 166* : 319-38.

Werblin, F.S. and Dowling, J.E. (1969) 'Organisation of the retina of mudpuppy *Necturus maculosus.* II Intracellular recording.' *Journal of Neurophysiology, 39* : 339-55.

Whitehead, C. and Bromage, N.R. (1980) 'Effects of constant long- and short-day photoperiods on the reproductive physiology and spawning of the rainbow trout.' *Journal of Endocrinology, 87* : 6P-7P.

Whitehead, C., Bromage, N.R. and Forster, J.R.M. (1978) 'Seasonal changes in the reproductive function of the rainbow trout *(Salmo gairdneri).' Journal of

Fish Biology, 12 : 601-8.

Whitehead, C., Bromage, N.R., Forster, J.R.M. and Matty, A.J. (1978) 'The effects of alterations in photoperiod on ovarian development and spawning time in the rainbow trout *(Salmo gairdneri).*' *Annales de biologie animale biochimie biophysique, 18* : 1035-43.

Whitehead, C., Bromage, N.R., Breton, B. and Matty, A.J. (1979) 'Effect of altered photoperiod on serum gonadotrophin and testosterone levels in male rainbow trout.' *Journal of Endocrinology, 81* : 139P-140P.

Wilhelmi, A.E., Pickford, G.E. and Sawyer, W.H. (1955) 'Initiation of the spawning reflex response in *Fundulus* by the administration of fish and mammalian neurohypophyseal preparations and synthetic oxytocin.' *Endocrinology, 57* : 243-52.

Winberg, M., Holmgren, S., and Nilsson, S. (1981) 'Effects of denervation and 6-hydroxydopamine on the activity of choline acetyl-transferase in the spleen of the cod, *Gadus morhua.*' *Comparative Biochemistry and Physiology, 69C* : 141-3.

Wisby, W.J. (1952) *Olfactory responses of fishes as related to parent stream behavior,* Unpublished Ph.D. Thesis, University of Wisconsin, Madison.

Wisby, W.J. and Hasler, A.D. (1954) 'The effect of olfactory occlusion on migration silver salmon *(O. kisutch).*' *Journal of the Fisheries Research Board of Canada, 11* : 472-8.

Witpaard, J. (1976) *Frog's vision: electrophysiological and developmental aspects,* Doctoral thesis, Leiden.

Wong, W.C. and Tan, C.K. (1978) 'Fine structure of the myenteric and submucous plexuses of a coral reef fish *(Chelmon rostratus).*' *Journal of Anatomy, 126* : 291-301.

Wood, C.M. (1975) 'A pharmacological analysis of the adrenergic and cholinergic mechanisms regulating branchial vascular resistance in the rainbow trout *(Salmo gairdneri).*' *Canadian Journal of Zoology, 53* : 1569-77.

Wood, C.M. (1977) 'Cholinergic mechanisms and the response to ATP in the systemic vasculature of the rainbow trout.' *Journal of Comparative Physiology, B. 122* : 325-45.

Wood, C.M. and Shelton, G. (1980a) 'Cardiovascular dynamics and adrenergic responses of the rainbow trout *in vivo.*' *Journal of Experimental Biology, 87* : 247-70.

Wootton, R.J. (1970) 'Aggression in the early phases of the reproductive cycle of the male three-spined stickleback *(Gasterosteus aculeatus).*' *Animal Behaviour, 18* : 740-46.

Wootton, R.J. (1976) *The biology of the sticklebacks,* Academic Press, New York.

Wotiz, H.H., Botticelli, C., Hisaw, F.L. and Bingler, L. (1958) 'Identification of estradiol-17β from dogfish ova *(Squalus suckleyi).*' *Journal of Biological Chemistry, 231* : 589-92.

Wotiz, H.H., Botticelli, C.R., Hisaw, F.L. Jr. and Olsen, A.G. (1960) 'Estradiol-17β, estrone and progesterone in the ovaries of dogfish *(Squalus suckleyi).*' *Proceedings of the National Academy of Sciences, U.S.A. 46* : 580-5.

Wourms, J.P. (1977) 'Reproduction and development in chondrichthyan fishes.' *American Zoologist, 17* : 379-410.

Yone, Y. (1978) 'The utilization of carbohydrates by fishes.' *Proceedings of the 7th Japan-Soviet Joint Symposium of Aquaculture, pp.* 39-48. Sept. 1978, Tokyo.

Young, J.Z. (1931) 'On the autonomic nervous system of the teleostean fish *Uranoscopus scaber.*' *Quarterly Journal of Microscopical Science, 74* : 491-535.

Young, J.Z. (1933) 'The autonomic nervous system of selachians.' *Quarterly Journal of Microscopical Science, 75* : 571-624.

Young, J.Z. (1936) 'The innervation and reaction to drugs of the viscera of teleostean fish.' *Proceedings of the Royal Society, B.120* : 303-18.
Young, J.Z. (1980a) 'Nervous control of stomach movements in dogfishes and rays.' *Journal of the Marine Biological Association, U.K. 60* : 1-17.
Young, J.Z. (1980b) 'Nervous control of gut movements in *Lophius*.' *Journal of the Marine Biological Association, U.K., 60* : 19-30.
Youngson, A.F. and Simpson, T.H. (1981) 'The effects of stream flow rate on the downstream migration of juvenile Atlantic salmon and on serum thyroxine levels in migrant fish.' *Third symposium on fish physiology*, Bangor, N. Wales, 1981: 26. (unpublished abstract).
Youson, J.H. (1982) 'The Kidney' in M.W. Hardisty, I.C. Potter (eds.), *The Biology of Lampreys, Vol. 3*, Academic Press, New York, pp. 191-261.
Zohar, Y. (1980) 'Dorsal aorta catheterisation in rainbow trout *(Salmo gairdneri)*. 1. Its validity in the study of blood gonadotrophin patterns.' *Reproduction Nutrition Développement, 20* : 1811-23.
Zucker, A. and Nishimura, H. (1981) 'Renal responses to vasoactive hormones in the aglomerular toadfish, *Opsanus tau.*' *General and Comparative Endocrinology, 43* : 1-9.

FISH INDEX

Fish are classified by families (Orders
in a few cases); (for superfamilies
referred to as — *oidea* in text, *see*
equivalent family, or cross-reference,
in index)

GENERAL INDEX